Hybrid Woodworking

BY MARC SPAGNUOLO,
The Wood Whisperer

The
Wood Whisperer
.com

Hybrid Woodworking

BY MARC SPAGNUOLO,
The Wood Whisperer

Hybrid Woodworking ■ Hybrid woodworkers use power tools for the grunt work of milling stock and roughing out parts. Then they turn to the workbench and hand tools to finesse the fit and finish. Hybrid woodworkers seek the best of both worlds.

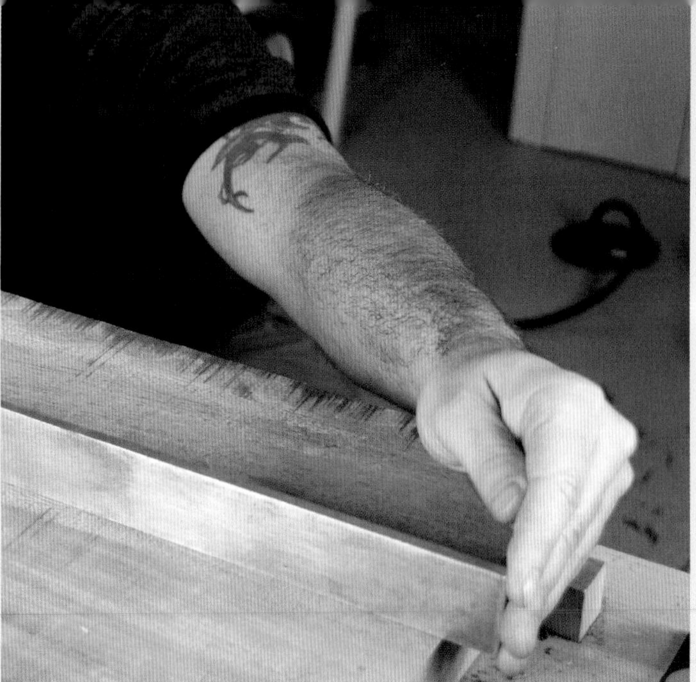

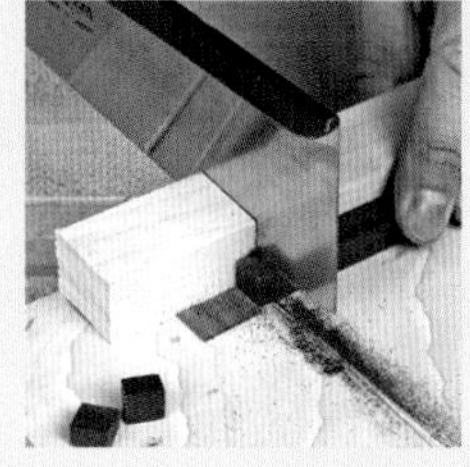

Contents

Dedication, Acknowledgments 6

About the Author 7

1 Introduction:
The Hand-Tool Renaissance 8

2 Tools of the Hybrid Woodworker:
Enhance, Don't Replace 16

The Basics: Power Tools and Machines 22
Must-Have Hand Tools 39
Hand Tools to Consider 75
Hand Tools to Consider…Maybe Later 84

3 Techniques of the Hybrid Woodworker:
Machines for Grunt Work, Hand Tools for Finessing 90

Milling Square Stock 93
Dados, Rabbets and Grooves 108
The Mortise-and-Tenon Joint 115
Half-Lap Joints 135
Dowels and Screw Plugs 142
Edge-Banding Plywood 144
Mortise for Butt Hinge 149
Hybrid Dovetails 154
Curves 169
Surface Preparation 175

4 Hybrid Woodworking Projects:
The Best of Both Worlds 178

Platform Bed 179
Split-Top Roubo Workbench 182
Wall-Hanging Cabinet 184
Adirondack Chair 186

Marc's Last Word 190

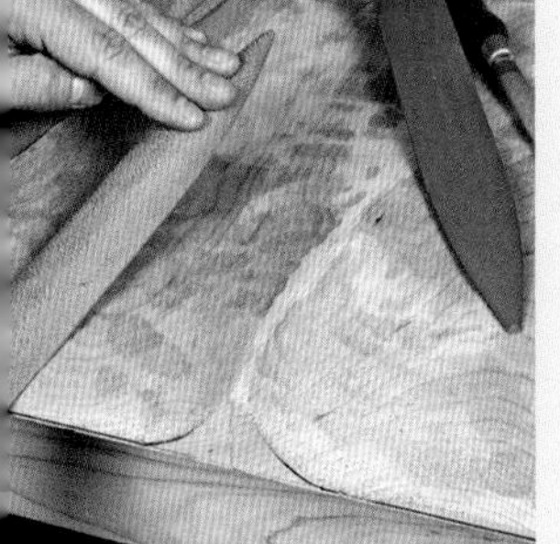

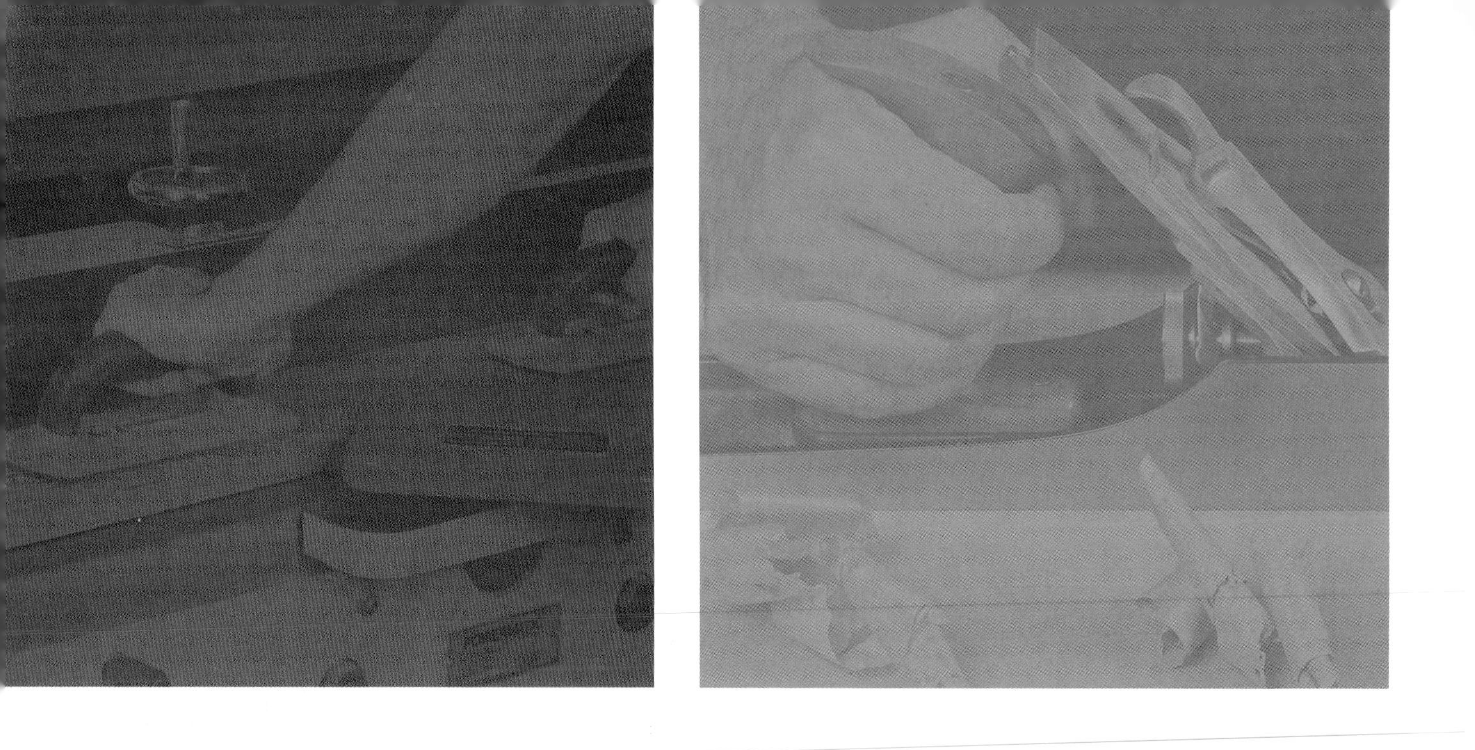

Dedication

This book is dedicated to my wife, Nicole, who constantly reminds me to follow my passion, and my son, Mateo, who constantly reminds me what to be passionate about.

Acknowledgments

I'd like to thank David J. Marks for his friendship, knowledge, and inspiration as well as Matt Vanderlist, Sam Vanderlist, Wilbur Pan, Shannon Rogers and John Kelsey for their assistance with a few technical aspects of the book. I'd also like to thank the online woodworking community for supporting my efforts since 2006.

Accessible Heroes ■ One of the great things about woodworkers is that they are accessible. I not only met but trained with my personal woodworking hero, David Marks (www.DJMarks.com).

About the Author

I cut my woodworking teeth on power tools, which isn't nearly as painful as it sounds. I was heavily influenced by TV shows such as *Woodworks* and *The New Yankee Workshop* and both shows displayed amazing power-hungry saws and sanders doing their work quickly and effectively. So it is no surprise that I modeled my tool collection after theirs; to do anything else just seemed illogical. It wasn't until a few years into the hobby that my hand-tool collection progressed beyond a beat-up old block plane and some dull chisels. Over time, I began to realize that hand tools, while unappealing to me as a complete system, had their uses. Slowly but surely, hand tools began filling voids left by my power tools. Card scrapers were doing the job of my random-orbit sander and a shoulder plane was allowing me to attain perfectly fitting tenons. Eventually, I realized that I had blurred the line between hand tools and power tools. I was practicing a hybrid of the two systems. Without realizing it, I had become a hybrid woodworker.

Even at the current stage of my woodworking career, my methods and preferences are constantly evolving. I can honestly say that my love of hand tools continues to grow so I am always looking for ways to incorporate new hand tools into my power-tool shop. You could say I'm just looking for excuses to buy new tools, but I prefer to see it as searching for ways to increase efficiency and reduce redundancies, while paying homage to the industrious woodworkers who laid the groundwork for our craft. To do this successfully, I need to remain flexible. I need to be willing to let go of any tool or technique, once I've become convinced that another tool or technique works better. That's why a necessary trait of all hybrid woodworkers is open-mindedness. We have to be willing to evaluate our current way of doing things and accept the fact that there could be something better, faster or easier. If you can't step outside of your comfort zone, this ever-evolving system of learning and adapting will never work for you.

My good friend David Marks, an accomplished woodworker and television personality, once described himself as a lifelong student of the craft. I have great respect for David and I can only hope to follow his footsteps and remain a lifelong student myself. I would encourage you to be a lifelong student and constantly challenge not only yourself, but also the status quo.

—Marc Spagnuolo, Phoenix, Ariz., June 2013

Hand Tools are Popular ▪ Hand tools are back baby! Whether shiny and new or old and well-used, more hand tools are finding their way into new woodworkers' shops every day.

Introduction: The Hand-Tool Renaissance

Have you noticed the most prominent and fastest-growing trend in woodworking these days? No, not midriff flannel shirts, silly! I'm talking about the return to our old-school roots.

We are experiencing a resurgence of interest in human-powered wood manipulation, a hand-tool renaissance. Every day more woodworkers are unplugging their power tools and choosing their traditional, non-electric analogs. Have you leafed through any of the current woodworking magazines lately? You may have noticed not only an increase in the number of hand-tool articles and reviews, but also the number of hand-tool advertisements. Do a few web searches and you'll find some popular blogs focused solely on hand-tool history, maintenance and use. In defiance of a hectic technology-driven world, it is clear that some of today's new woodworkers have the desire to hit the pause button and take some time to carve their own paths.

We resist the urge to stare into screens all day and instead retreat to the workshop to craft functional and beautiful creations with our own hands. Working wood can take some time and a woodworker's world seems to move at a slower pace than the rest of the population's. For us, the journey is as important, and sometimes more important, than the destination. So is it any wonder that when given the choice between a power tool and a hand tool, many woodworkers are drawn to the more modest option?

This is a very good thing for the craft. Hand-tool techniques are the foundation for the jobs that power tools do. The powered jointer is just an upside-down bench plane with a motorized blade. A band saw is the modern equivalent of the traditional frame saw. Sanders do the work previously assigned to smoothing planes. Our new tools and our old ones are so interconnected that it certainly couldn't hurt to have a strong working knowledge of hand tools, even if you intend to focus on power-tool woodworking. Understanding their hand-tool ancestors helps us get the most out of our power tools, and in some cases can help us identify where a modern power tool falls short and how we could overcome such limitations.

The benefits include increased gratification, a higher degree of craftsmanship and a more pleasant and productive woodworking environment.

Substantial Benefits

Adding hand tools to a machine woodworking shop brings several substantial benefits:

- Hand tools are quieter.
- Hand tools are safer.
- Hand tools produce less dust.
- Hand tools improve accuracy.
- Hand tools can save you money.
- Hand tool projects have a special quality about them.
- Hand tools can be more gratifying.

I am a new parent, and I'm very thankful that my hybrid woodworking routines allow me to get some work done while my son is sleeping. If your day job means you can only woodwork at night, or if you live in an apartment, the quiet nature of hand tools could make the difference between woodworking and not woodworking. The sound of a handsaw, the mallet taps on chisel, and the swoosh of a handplane are the only sounds that will emanate from your shop. Plus you have the added bonus of actually being able to hear the radio, or your own thoughts, or in my case, the baby monitor.

Hand tools are generally safer to use than power tools. Sure, you can still end up in the emergency room if you aren't careful, but injuries with hand tools very seldom result in lost digits. And, great news for your lungs, the by-products of hand tools are shavings and chips that you can sweep up, not fine dust you'll be inhaling and coughing up.

Hand tools improve accuracy because they remove small, controlled amounts of material with each stroke or pass. You can sneak up on the perfect fit. And should you go a stroke or two too far, you probably haven't ruined the workpiece. Go a pass too fast or too far with a power tool and you'll be starting over.

Hand tools also offer a potential economic benefit. By using scrapers and planes, you'll consume less sandpaper. Good quality sandpaper doesn't come cheap, and with a hybrid approach you'll use much less of it and you'll likely need only to stock up on one or two grits. Additionally, because you aren't producing nearly as much dust, you won't have to purchase as many of those expensive replacement bags for your dust extractor or shop vac. When your waste products are shavings and chips, a broom and dust pan becomes a viable cleaning solution.

Also, keep in mind that one of the great joys of hybrid woodworking is the ability to sneak up on the perfect fit. To facilitate this, most power-tool cuts will leave the workpieces oversized. So the chances of cutting a piece too short or making a tenon too small are minimized. That means less waste and fewer do-overs.

The final benefit to consider is the big X-factor, gratification. How much personal satisfaction you get from using hand tools is, well, personal. But I believe we all have the capacity to appreciate the magical feeling of achieving a positive result with a finely-tuned hand tool. And as much as I love to see great results from power tools, I have to admit that a high-quality hand-tool result is a little more special.

Based on all of these benefits, learning about and incorporating hand tools into the wood shop certainly sounds like a great idea. But don't get me wrong. I love my woodworking machines and my portable power tools, and I have no desire to become an exclusive hand-tool user. But I find there are a lot of good reasons to strategically incorporate hand tools into my power-tool shop.

So why doesn't every woodworker hold a ticket for a ride on the hand-tool train? To understand why, we'll need to consider the path of the modern woodworker. As popular as hand tools are, the current landscape still presents numerous obstacles, pitfalls and diversions that create confusion and can send the information-seeking woodworker down a rabbit hole.

A Diamond in the Rough? ■ In spite of a coat of rust, some old flea-market finds can be rehabilitated into functional working tools.

Power Tools First, but Why?

Today's new woodworker tends to follow a typical path involving the acquisition of basic power tools (table saw, drill, router, circular saw, jigsaw, and sander). I chat with thousands of woodworkers every year and I have the good fortune of hearing many origin stories about how they entered the craft. Some took shop class in high school, some recall lazy Sundays in the garage with grandpa, but most were influenced by a TV show or inspired after doing their first DIY or honey-do project. One thing nearly all of these people have in common is that their first tools are powered. This is actually the opposite of how it should be. In just about any discipline, one should always begin with the rudimentary tools first, and work up to the more complex gear. These days, that rarely happens.

So why do so many people start off with power tools first? I think they get the woodworking bug from watching DIY home-improvement shows. These shows are aimed at a general audience and they tend to the viewpoint that hand tools are outdated relics. So it's no surprise that many new woodworkers' first tool purchases will be 100-percent powered and they will start out with no appreciation for traditional hand-tool woodworking.

For example, *The New Yankee Workshop* ran on PBS for 20 years from 1989 to 2009 and influenced countless people, including yours truly, to become woodworkers. The host, Norm Abram, became a celebrity woodworker and he is widely perceived as a power-tool junkie. I am certain Norm has a strong working knowledge of hand tools, but he rarely used them on the show. Perhaps the power-tool sponsors played a role, or perhaps that was what the producers believed the audience wanted. Either way, a new woodworker would walk away with the impression that hand tools just aren't used any more.

I love my woodworking machines and my portable power tools, and I have no desire to become an exclusive hand-tool user.

Planting the Hand-Tool Seed

Regardless of how people enter the craft, many will eventually become acutely aware of hand tools. Even if they initially regard hand tools with apprehension, most woodworkers develop

Good Form ▪ Take a wide stance while planing, letting your body move the plane across the board, not your arms.

an appreciation for what these tools can do in skilled hands. As more and more woodworkers turn to the Internet, they are further exposed to blogs, podcasts, and forums where the hand-tool influence is alive and well. And if they begin subscribing to popular woodworking magazines, in most cases they'll find a well-balanced presentation of both hand- and power-tool use. One way or another, the hand-tool seed gets planted. The question is, will it be allowed to grow?

Unfortunately, because most woodworkers already have a decent complement of power tools, adding hand tools can be expensive. In some extreme cases, woodworkers vow to replace all of their tools with hand-tool equivalents, which is, in my opinion, a big mistake. If you take the hand-tool plunge too soon and too deep, you might find yourself overwhelmed with this new paradigm. Other folks take a more measured approach, adding a new hand tool here or there and discovering over time by trial and error which tools are actually useful. Regardless of how the hand tools find their way into the shop, as an invested power-tool user, money is going to be a major factor. After I made my initial investment in a power-tool shop, I had a tough time persuading the finance committee that I really needed more hand tools. And frankly, I had a tough time convincing myself it was the right thing to do.

To make matters worse, hand tools are no longer the inexpensive alternative they once were. A decade ago, you could find decent fixer-uppers on eBay or at a local flea market. Today, eBay is overpriced and extremely competitive, and good luck finding an undiscovered box of tools at a flea market that someone hasn't already pilfered. Good deals can still be found, but you're going to work hard to get them. And if you want to purchase a new hand tool that doesn't need much in the way of tune-up or fettling, the price can range from merely expensive all the way to ludicrous.

The Challenge of Maintenance

Now let's assume the new woodworker overcomes these obstacles of power-tool influence, expense and pride, and acquires a basic complement of usable hand tools. The next challenge is the fact that hand tools require precise care and maintenance. Older tools may need complete rehabilitation including rust removal, sole flattening, filing, japanning and blade replacement. High-quality new tools still require basic maintenance including back-flattening and blade honing as well as general fine-tuning for particular tasks. Even if you are accustomed to the many maintenance requirements of power tools, hand tool fine-tuning and maintenance can be a rude awakening. Thankfully, most woodworkers are not easily deterred by challenges and those who press on will eventually develop a working knowledge of how to maintain and tune their hand tools. Like any other skill, it takes practice.

The Challenge of Use

The next challenge is learning how to use the tools, which really should be the fun part, right? In some cases it is, but there are many frustrated woodworkers who simply cannot get good results from their hand tools. Some become so disheartened that they begin to consider giving up the craft altogether. The truth is, wielding hand tools is a lot like hitting a golf ball. The best club in the world won't do anything for you if you don't know how to swing it. Getting a hand tool to work effectively takes practice and patience. And just like the golf swing, the best thing you can do is have someone critique your form by watching you work. Stance, posture, arm and leg movements, hand placement and grip are all major concerns to the hand-tool user. Someone coming from the power-tool world is liable to underestimate the nuance involved in effectively passing a hand tool over a piece of wood. The hand-tool learning curve is steeper than that of most power tools.

Ultimately, woodworkers who jump into the wild and crazy world of hand tools too fast, too soon and without proper guidance may end up discouraged and soured on the craft. A crappy untuned hand tool in the hands of a novice won't just produce lackluster results, it won't produce any useful result. It will contribute to the firewood pile while slowly driving its owner into dark madness. Who wants to live in a world filled with torn-out grain, undulating surfaces, non-square edges, calluses and sore forearms?

Don't get me wrong, the picture is not entirely grim. Hand tools are gaining wide acceptance and we are incredibly lucky to have an over-abundance of free and paid resources to help guide us on our woodworking journey. Thanks to books, DVDs, blogs, podcasts, online courses, local seminars and woodworking schools, there has never been a better time to learn the craft of woodworking with hand tools. At the same time, many of these resources don't consider the possibility that not everyone wants to use hand tools exclusively.

Most woodworkers would never willingly give up their power tools, yet many still have

Shoulder Maintenance ▪ A tenon cut at the table saw can be fine-tuned at the workbench.

a growing interest in what hand tools might have to offer. The risk is becoming persuaded to purchase a whole host of hand tools they may never wind up using. It wouldn't make sense to invest in a drop-dead gorgeous Lie-Nielsen scrub plane if you never intend to flatten a rough board by hand. Discussing hand tools in a vacuum with no consideration for their power-tool equivalents inevitably leads to significant

A Broader Perspective

Now it's time for a little disclaimer. There will likely be several times in this book when you might wonder, "Hey, wouldn't it be quicker to do XYZ instead?" If that question pops into your head, I urge you to consider the reasons you work wood in the first place. Is it always about expediency or do you subscribe to the romantic notion that we are doing more than simply building furniture?

As I see it, we are taking one of nature's most amazing natural resources and turning it into a functional and beautiful "something" that can be utilized, appreciated and enjoyed for years well beyond our own. While there are certainly limits to the amount of time I want to spend doing mundane woodworking tasks, having the opportunity to fine-tune joints prior to assembly is a gratifying treat that makes me feel closer to my work and gives me higher quality results. If it happens to eat up a few more minutes on the shop clock, and many times it will, I really don't mind. My work and my mental health both benefit from the extra care and attention.

At the same time, I do have my limits. No one on this planet can convince me it's in my best interest to mill a board 4-square by hand, but that's just me.

redundancies. For instance, if I plan to continue using my powered jointer, I have little need of a No. 7 jointer plane. For those who like to collect tools, this is no problem. But for the budget-conscience woodworker, redundancy in tooling can be wasteful, frustrating and confusing.

In my opinion, the discussion is not about hand tools versus power tools as if they were two mutually exclusive categories. Instead, let's talk about how to enhance our existing shops by incorporating carefully chosen hand tools. A similarly useful discussion might center around how to enhance a hand-tool shop with carefully chosen power tools. Because that isn't the kind of shop I run, I can't offer much perspective. I have a well outfitted power-tool shop, and I am most comfortable teaching people how to improve the accuracy, speed and quality of their work with strategic hand-tool selection and use. In fact, that's the exact purpose of this book: to show you how hand tools and power tools can be used together in an efficient, harmonious and gratifying balance. By rights, this all falls under the broad label of woodworking. But because hand tools and power tools have become somewhat segregated in the hearts and minds of many woodworkers, I like to call this best-of-both-worlds system "hybrid woodworking."

Let the power tools do the grunt work and the hand tools do the finesse work.

Hybrid Woodworking

The goal is to use the best tools for the job. The basic tenet of the hybrid woodworker is to let the power tools do the grunt work and the hand tools do the finesse work. In years past, woodworkers had apprentices to do the grunt work, leaving the finer details to the craftsmen who were higher up the chain of command. Today I work alone and my apprentice is a toddler, but I do have help with the grunt work from my power tools. Whether it's milling boards flat or hogging away excess wood for some particular joint, my power tool "apprentices" allow me to get the job done quickly and efficiently so I can take my time with the details that matter.

As a quick example, I usually batch out tenons using a dado blade in the table saw. It's a fast and repeatable way to make the joint. But the saw leaves a rough tenon cheek and it can be difficult to set the blade at the perfect height, so I leave tenons a bit oversized. I then head over to the workbench and finesse the cheeks using my shoulder plane, rabbeting block plane or router plane. By removing super-thin shavings I can creep up on the perfect fit while also making the tenon nice and smooth. If I go one stroke too far, I'm only a few thousandths of an inch past my desired outcome. If I were doing this finesse work at the table saw, any error would be much greater. Not only does this methodology give me better results and tighter joints, it also avoids major errors. And on a personal level, the process itself just makes me feel good about what I do.

Why Doesn't Everyone Use The Hybrid System?

In my opinion, the hybrid approach is the most flexible, efficient, accurate and gratifying way to work wood. So why doesn't everybody use this system? Well, some hand-tool devotees are doing woodworking simply because they want to take their time and enjoy a process that has its roots in a bygone era. Perhaps they have a stressful job or a hectic personal life, and woodworking provides a relaxing break. These folks certainly aren't naive to the fact that they could use a router to create a mortise, but chopping that mortise with a chisel and a mallet gives them a more intimate relationship with the furniture and the craft as a whole. To each his or her own.

While some believe power-tool woodworkers are largely focused on the destination and not the journey, that isn't true for me. I enjoy every second of my time sculpting curves with a grinder, pulling glass-smooth figured boards out of my 15" helical-head planer and slicing thin veneers with a powerful band saw. To me, the roar of a powerful motor is the sound of a bird singing a delightful tune. Just as a hand-tool

Neanderthals vs Normites

Sometimes, especially on the Internet, woodworkers are unfairly lumped into two broad categories: hand-tool users and power-tool users. There are fanatics on both sides of the fence who are at the root of this phenomenon. Some hand-tool zealots believe that unless you completely unplug, you're not a real hand-tool user. Some power-tool proponents believe that hand tools are just a waste of time and represent an inefficient step backward. This fanaticism typically generates a counter-reaction in both camps where folks who favor one thing or the other feel the need to defend their choices and the gap only gets wider.

The online woodworking community has been using pet names for years to identify these seemingly disparate groups: Neanderthals and Normites. The Neanderthals are the knuckle-dragging hand-tool lovers who seem to enjoy slow grunt work. Normites, named after the self-confessed power-tool junkie Norm Abram, won't use a tool unless it has a power cord and would rather spend an hour setting up a dovetail jig than 10 minutes cutting the joint by hand.

Of course, I'm having fun at the expense of the extremists because I believe that most folks are better off somewhere in the middle and have a natural desire to be there. But the current state of the online woodworking world compels new woodworkers to choose a side as if hand tools and power tools are mutually exclusive. This is unfortunate because there is tremendous value in knowing how to use all woodworking tools.

Ultimately, you should do the kind of woodworking that makes you happy. But if you're new to the craft, I would hate to see you go down one particular path at the expense of the other. The truth is, both systems have strengths and weaknesses. Fortunately, where many power tools sometimes fall short, hand tools excel. And conversely, where hand tools fall short, power tools excel. That's why, in my humble opinion, hybrid woodworking is the least resistant path to woodworking greatness, as we truly enjoy the best of both worlds.

devotee gets a charge out of making beautiful slivers with a finely honed chisel, the power-tool devotee gets amped up when a 5hp motor spins a carbide-tipped blade through 12/4 maple.

I think all woodworkers are here for the same fundamental reasons. We all have at least some appreciation for each step of the woodworking process: selecting the wood, milling, joining, shaping, assembling and finishing. The tools we use to work our way through these steps are just that: tools. Whether using power tools, hand tools or a hybrid combination, we're playing for the same team. After all, isn't everybody sick of mass-produced particleboard furniture? I know I am.

When it's all said and done, the recipient of your craftsmanship and all future onlookers will have no idea what tools you used to make it, nor will they care. So the vehicle you choose for your particular journey from pile of boards to masterpiece is completely and totally a personal one. Everyone does whatever makes them happy, in the workshop and in life. I hope you'll consider the benefits of the hybrid woodworking approach because I truly feel it makes the whole process more enjoyable and keeps the door open for new ideas and new methods of work.

About this Book

This book covers hybrid woodworking basics. I'll discuss the essential tools every hybrid woodworker should own, as well as which ones you might consider. I'll demonstrate numerous woodworking joints and show you how to approach them from a hybrid perspective. If you are new to the craft and haven't started accumulating tools yet, by the end of this book you'll have a much better understanding of what tools you need for the things you want to make.

We'll conclude the book with a brief overview of projects I've made in the past, highlighting areas where specific hybrid techniques were used.

2

Tools of the Hybrid Woodworker: Enhance, Don't Replace

Redundancy = Options ■ While the shoulder plane, rabbeting block plane and block plane are very different tools, they do share some functionality.

Have you ever heard anyone claim that hand tools are slow? Perhaps this is a notion you currently subscribe to. If so, you're not alone. In fact I'd guess that this is the primary reason why most power tool enthusiasts don't even consider hand tools as a viable option in the shop. At this point, I bet you're expecting me to try to convince you that hand tools are not slow, but I'm not going to do that. Truth be told, I agree with the notion, but only partially. In my experience, hand tools are only slow when I try to use them as replacements for the jobs currently done by my power tools.

I don't think anyone can argue against the fact that milling a board by hand using several bench planes and winding sticks will certainly take longer than using a powered jointer, planer and table saw. Ripping a board to width at the table saw is, without a doubt, faster than using a handsaw and a sawbench. So one basic rule of the hybrid woodworking method, as far as it concerns tool buying, is "Enhance, don't replace." Don't trade your sliding compound miter saw for a miter box. Don't sell your band saw for a bowsaw. And please, for the love of all things made of wood, don't trade your power planer for a scrub plane. I know several hand-tool zealots who still keep a power planer in the shop for easy milling when no one else is watching.

Keeping the "Enhance, don't replace" rule in mind, a hybrid woodworker's tool box begins to take shape, and it probably looks different than you might have anticipated. Most notably, various handplanes and saws will be conspicuously absent. Remember, we are trying to prevent excessive functional redundancy. So what we look for are specific tools that perform specific functions: the ones our power tools either aren't capable of

Clean Your Cheeks ▪ A rabbeting block plane excels at cleaning up the cheeks of a tenon.

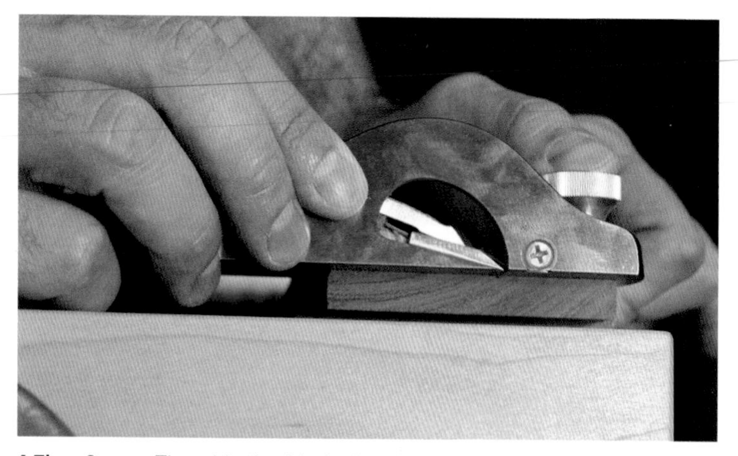

A Time-Saver ▪ The rabbeting block plane features a full-width blade that makes finessing a tenon quick and easy.

or simply don't do with as much precision and accuracy as we'd like.

In addition to the "Enhance, don't replace" rule, there are a few other guidelines we can apply to help decide if a particular hand tool should become part of our collections. If a tool can bring one or more of the following benefits to the table, then it should at least be considered as a potential purchase:

- functionality
- efficiency
- accuracy
- or gratification.

Keep in mind that there is a lot of room here for personal preference and no two tool kits need to look exactly the same, but evaluating these potential benefits and asking yourself the associated questions will allow you to make a more objective decision about every tool that enters your shop, powered or not.

Decision-Making Process

Let's use an example to illustrate the question-and-answer process. I'll also include a score that will help you derive a numerical rating for how well the tool satisfies each benefit. Here's the scenario: I already own a medium shoulder plane and a block plane. Should I purchase a rabbeting block plane?

1. Functionality

Does the new tool bring something unique to the table and is it a multi-tasker?

While there is overlap between a medium shoulder plane and a rabbeting block plane, there also are some important differences. A shoulder plane can be used to clean up not only the shoulders of a tenon but also the cheeks. Unfortunately though, a medium shoulder plane is quite narrow and would require multiple passes on the tenon cheek. This could result in unevenness and gouges on the tenon cheek, as well as inconsistent tenon thickness. A rabbeting block plane is significantly wider and covers more real estate per pass. That makes it the better choice for smoothing a tenon cheek. My standard block plane would cover just as much real estate per pass, but its blade does not extend all the way to the sides of the plane body, which means the cut will stop short of the tenon shoulder, which is certainly not ideal. So in my opinion, a rabbeting block plane does offer some additional functionality over both the shoulder plane and the block plane.

A rabbeting block plane isn't just helpful with tenons. You can use it to tune up other joints as well, and as the name suggests, it can do many of the jobs that a standard block plane can do. In fact, if you don't already own a block plane, I feel a rabbeting block plane is the better first purchase. But because I already have a standard block plane, the rabbeting block plane does bring some redundancy. So I'll need to decide if its additional functionality justifies bringing it into the fold.

In terms of functionality for my particular situation, I give the rabbeting block plane a score of 2 out of 5. If I didn't already have a shoulder plane and a standard block plane, this score would be higher.

2. Efficiency

Will the tool make specific processes more efficient?

When finessing the fit of tenons, neither the shoulder plane nor the block plane are ideal for trimming tenon cheeks. The rabbeting block plane, however, will make quick work of the task. Fewer passes on each tenon face means the job gets done faster and I can move on to the next part of my project. Additionally, some models (such as the one sold by Lie-Nielsen) feature a nicker that slices the grain ahead of the blade, helping to prevent tear-out and resulting in a cleaner cut.

In terms of efficiency, I give the rabbeting block plane a score of 4 out of 5. To illustrate why, let's look at some actual times. When using a shoulder plane, it takes me about two minutes to clean up a typical tenon, including time for a couple of test fits. With the rabbeting block plane I can do the same work in about half the time, thanks to the wider body and full-width blade. For a set of four frame-and-panel doors (16 tenons), it would take me about 32 minutes using the shoulder plane and 16 minutes using the rabbeting block plane. This is just a ballpark estimate but it does illustrate that over the course of a full project, small time savings can really add up.

3. Accuracy

Will the tool increase my accuracy?

It comes down to making fewer passes to get the job done. Fewer passes translates to more consistent results and fewer opportunities for error. When you use a plane to clean up a tenon cheek, it's important to count your strokes so that you can repeat the same number of passes on the other side. If you don't do this, the tenon becomes off-center. Reducing the number of passes makes it easier for me to count, remember and execute the proper number of passes on both sides of the tenon.

In terms of accuracy, I give the rabbeting block plane a score of 3 out of 5. While fewer passes does increase my overall accuracy, I can get good results with my shoulder plane if I don't lose count and I practice solid technique. So there is an increase in accuracy with the rabbeting block plane, but it isn't dramatic.

4. Gratification

Will the tool add to my enjoyment of woodworking?

Without a doubt, gratification is the wild card. If you work wood purely for the enjoyment of the process, this benefit could easily trump the other three. Let's say a particular tool scores very low in the previous three benefits, but you

Shoulder Plane ▪ The shoulder plane, as the name implies, excels at finessing end grain on a tenon shoulder.

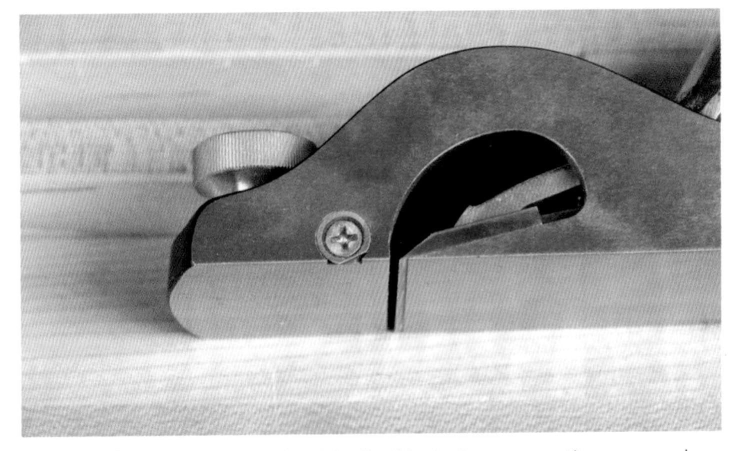

The Nicker ▪ The nicker on this rabbeting block plane severs the cross-grain fibers ahead of the blade to yield clean cross-grain cuts.

Item/Tool Name:

What purpose(s) does this item serve?

☐ Milling ☐ Carving/Sculpting ☐ Shop Organization
☐ Measuring/Marking ☐ Assembly ☐ Safety Equipment
☐ Joinery ☐ Finishing
☐ Other

Does this item serve a similar purpose to others that I have?

☐ No ☐ Yes, it's similar to

Do I need this item? (Check all that apply)

☐ I can't build a particular project without it ☐ I want this to work more safely
☐ I want this to build faster ☐ I want this for convenience
☐ I want this to increase build quality/accuracy ☐ I just want it

Category	Score (out of 5)	Multiplier	Total
Functionality	/ 5		
Efficiency	/ 5		
Accuracy	/ 5		
Gratification	/ 5		
		Grand total score	/ 100
		Purchase threshold	
		Should I buy this?	☐ Yes ☐ No

Notes:

Item/Tool Name:
Rabbeting Block Plane

What purpose(s) does this item serve?

☐ Milling ☐ Carving/Sculpting ☐ Shop Organization
☐ Measuring/Marking ☐ Assembly ☐ Safety Equipment
☒ Joinery ☐ Finishing
☐ Other

Does this item serve a similar purpose to others that I have?

☐ No ☒ Yes, it's similar to Medium shoulder plane and Block plane.

Do I need this item? (Check all that apply)

☐ I can't build a particular project without it ☐ I want this to work more safely
☒ I want this to build faster ☐ I want this for convenience
☒ I want this to increase build quality/accuracy ☐ I just want it

Category	Score (out of 5)	Multiplier	Total
Functionality	2 / 5	5	10
Efficiency	4 / 5	5	20
Accuracy	3 / 5	5	15
Gratification	3 / 5	5	15
		Grand total score	60 / 100
		Purchase threshold	50
		Should I buy this?	☒ Yes ☐ No

Notes: If I didn't already have a shoulder plane and a block plane, this would be a no-brainer. But it still exceeds my purchase threshold. Off to Lie-nielsen.com!

absolutely love the process of using it. I don't know, maybe you just like the way you look holding it? I know for me personally, I enjoy cutting mortises with a router and then squaring the corners with a chisel. Could I have used the hollow-chisel mortiser instead? Certainly. But there's something gratifying about the process of squaring things up with a chisel and mallet. Ultimately, if you value a new tool's potential for gratification more than the other benefits, a low-scoring tool may still find its way into your collection.

In terms of gratification, I give the rabbeting block plane a score of 3 out of 5. A rabbeting block plane doesn't do much to increase my personal enjoyment of the process, although being more efficient and a little more accurate always makes me happy. Plus I find it more comfortable in the hand than comparable tools such as the shoulder plane.

A Personal Process

As you can see, this can be a very personal process. Flexibility relies on comparisons to your unique tool set. Accuracy and efficiency depend not only on your tool set, but also on your level of experience and skill. Just because you're a woodworking ninja with a handsaw doesn't mean I can wield the tool with the same level of skill. And gratification is 100 percent personal. After all, while many woodworkers find it satisfying to chop a mortise by hand, I'd find it more pleasurable to chop a mortise in my hand. That's an exaggeration, but you get the point: We all have different goals, preferences and skills.

Going through this Q&A scoring exercise allows you to think critically about each benefit and how useful the tool is likely to be. If you want a systematic way to approach tool-buying decisions, you can assign weights to each benefit. Each benefit rating then contributes to the total point count and you'll end up with a simple buy/don't buy score. Above is a sample worksheet that gives each benefit equal weight. In other words, equal weight means that I feel all four benefits are equally important to me.

Functionality 2/5, efficiency 4/5, accuracy 3/5, gratification 3/5, using a multiplier of 5, my total score would be 60 out of a possible 100 points. If I set my purchase threshold at 50, any tool that receives a score of at least 50 points will find its way into my shop. In this case, with a score of 60, the rabbeting block plane is victorious and will soon find itself nestled comfortably in my tool chest.

Customizing the Calculations

Everyone has a different way of seeing things. A tool that I rate as 3/5 might score 2/5 in your estimation. You might find that your purchasing threshold is higher than 50. And you also might find that you value one of the ratings categories higher than the others. To get the most out of this tool-buying decision matrix, you need to dig a bit deeper and customize the calculations for your particular situation.

The purchasing threshold is a personal decision. Thankfully, finding yours should be pretty easy. I've found it helpful to run through this exercise on a few tools I already own. Compare the scores for tools you use a lot and enjoy immensely to others that don't get used all that often. This should give you a solid frame of reference and a range from the high scoring no-brainer purchase to a low-scoring tool you'd skip buying. This simple exercise will greatly improve the effectiveness of the worksheet. If you use my purchase threshold of 50 points without giving it any more thought, you might find yourself missing out on tools you should buy or, worse yet, buying tools you don't need.

To further customize the test to your personal situation, you can modify the multiplier. You probably noticed that by default, I use a multiplier of 5 for each category. This facilitates easy calculations by putting the scores on a simple 100-point scale. One alternative would be to rate each category on a scale of 25 points, which I'd find very difficult to do in any meaningful way. For me, a range of 1 to 5 requires a lot less in the way of mental gymnastics. Multiplying each score by 5 puts us on the 100 point scale and gives each category equal weight. It's like saying functionality, efficiency, accuracy and gratification are all equally important to me. But what if I'm a happy retiree who, perhaps, feels that gratification is more important than efficiency? I can simply increase the gratification multiplier to 6 and decrease the efficiency multiplier to 4. If I run a tool-purchase test and both gratification and efficiency receive the same score of 2/5, gratification will contribute 12 points to the total while efficiency only contributes 8 points. After a little soul searching you can fine-tune this test so that it is accurate and effective for you. Keep in mind you can modify the multipliers as much as you like, as long as the highest possible point total is 100. So if you raise the multiplier for one category, you need to decrease the multiplier for another category by the same amount.

Not everyone will want to get this nerdy about their tool choices, but sometimes it really helps to break things down in the most objective way possible. With the help of a little math, deciding whether or not to buy a tool will no longer be a long and drawn-out experience. Whether you actually want to spend the money is a whole other issue.

My Recommendations

Now that you have a decision matrix for future tool-buying decisions, I hope to save you a little time by giving you some basic tool recommendations. Although this book does assume you're somewhat familiar with the tools found in a typical power-tool shop, we'll take the time to review what I feel are the core power tools that are essential to the milling and joint-making processes. We'll then turn our attention to the hand-tool world, focusing on the must-haves first. These are the tools I feel no hybrid woodworker should be without. Because every person is different, my kit also will contain some tools that might or might not be useful to you, depending on your circumstances and preferences. Tools that fit this profile are what I think of as tools to consider. Finally, some hand tools simply are not needed in the hybrid woodworking shop and those are regarded as tools to avoid, at least for now.

My tool recommendations are by no means exhaustive or absolute. My suggestions are based on my personal experience and the workflow I've adopted as a custom furniture maker. I suggest using these lists as a starting point and foundation. From there, you should be able to begin carving your own path to woodworking success. When it's all said and done, your tool kit might look very different from mine. You also might find that your tool kit morphs over time as your needs change and you yourself evolve as a craftsperson. My goal is to make sure you understand which tools are useful and why, as well as which tools represent excessive redundancies and are better off left on the store shelf.

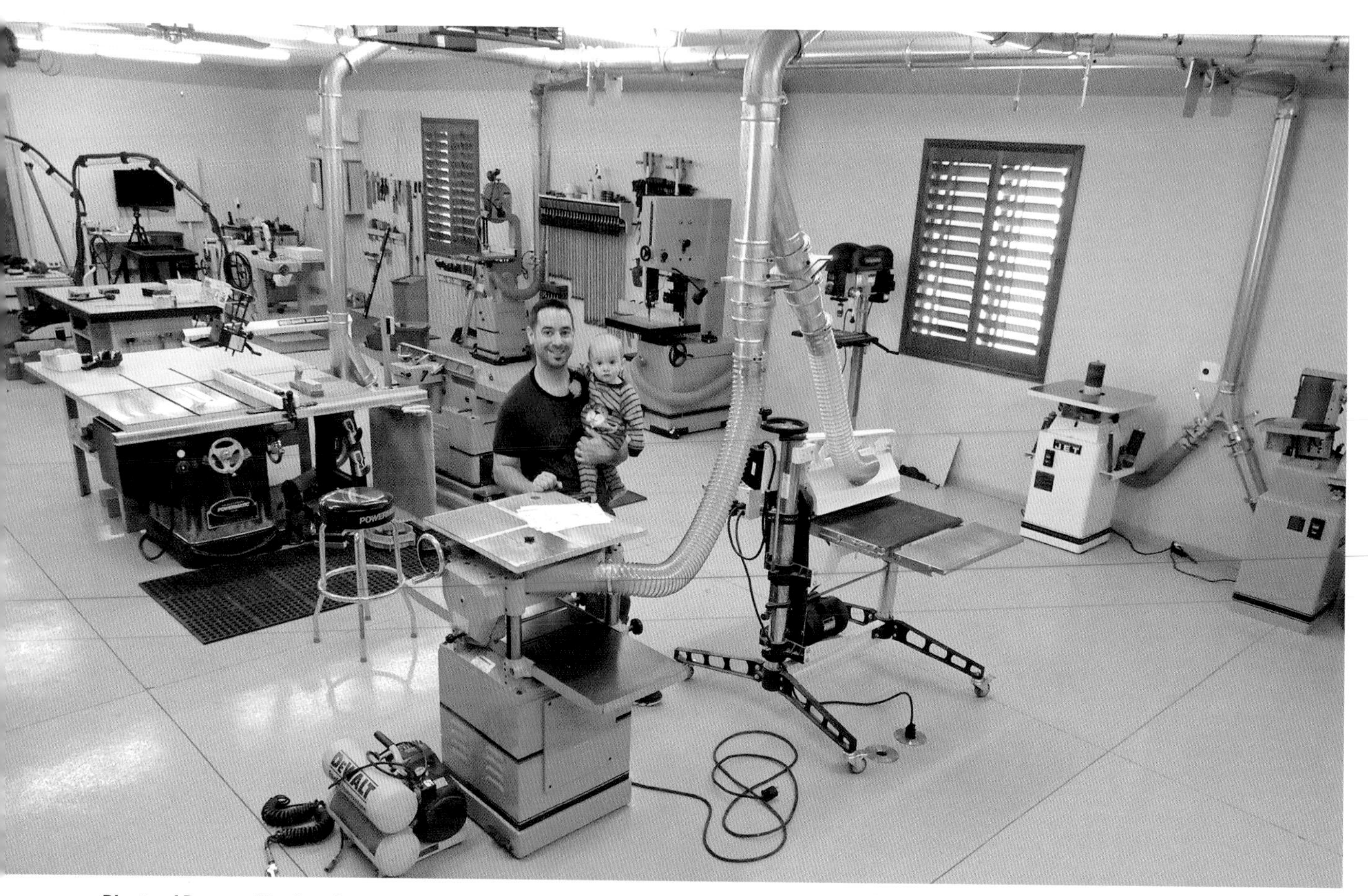

Plenty of Power ■ My shop features all of the typical stationary power tools. Toward the front of the shop, beyond the table saw, is my workbench and hand-tool area, which takes up substantially less space.

The Basics: Power Tools and Machines

Before discussing each power tool individually, I want to give you a brief overview of how I process project stock from rough boards to finished piece. This will give you the context surrounding each tool and why I consider it vital to my success.

I use a jigsaw to break rough wood down to rough length. From there, if necessary, I use the band saw to rip large boards to rough width. I go to the jointer to clean up one face and one edge of the board. The thickness planer next makes the other face parallel to the jointed face while it brings the board to its desired thickness. Now to the table saw to rip the board to final width. Next I use either the miter saw or the table saw to trim the boards to final length. From there, the process becomes project-dependent. Typically, I rout mortises but rely on the table saw to handle just about any other primary joint-making task. If the project involves plywood or other sheet goods, I break the 4x8 sheets down to manageable size using a portable circular saw. Then I go to the table saw to trim the pieces to final length and width as well as to cut any necessary dados, rabbets and grooves.

In this system, hand tools typically come into play when sweetening joints to a perfect fit, and smoothing surfaces before finishing. Along with increased accuracy and better final surfaces, hand tools allow me to avoid most of the noisy and dusty sanding that's typically done in the power-tool workshop. My hand-powered scrapers and planes create such a fine surface that all I need do to make it ready for finishing is

Heart of the Workshop? ■ The table saw is a powerful and versatile multi-tasker. Mine has an accurate fence system, auxiliary tables for managing large stock and a great paint job.

a once-over sanding with #220-grit paper on the random-orbit sander.

Of course there are more tools outside this general workflow, but the ones I plan to focus on here relate specifically to milling, joining and finish preparation. This is not intended as a complete workshop inventory. At the end of this section, I include a list of additional tools I feel are important, although they don't necessarily fit into this hybrid woodworking conversation.

Work Safe! ■ A table saw can be terribly dangerous if used improperly. Be sure to use task-appropriate safety gear such as guards, riving knives, push sticks, push blocks and featherboards.

Table Saw

It's been said many times that the table saw is the heart of the workshop. That may not be true for all woodworkers but I don't think anyone can deny the tool's versatility. Unfortunately, when used in an unsafe manner, the tool can create absolute mayhem in milliseconds. When used with a riving knife or splitter, a blade guard, a task-appropriate push stick and a healthy dose of respect, the table saw is a fantastic addition to any hybrid woodworking shop.

Miter Gauge ■ An after-market miter gauge adds functionality and accuracy to the table saw for making simple crosscuts, miters and an array of complex joints.

How I Use the Table Saw

Using nothing more than the stock fence, the table saw can perform numerous critical operations including ripping boards and sheet stock to width and cutting such joints as dados, rabbets and grooves. Add the stock miter gauge and you can crosscut boards and sheet stock to length and create such joints as miters, tenons, half-laps and many more. Use the built-in blade tilt and you can saw bevels from 0 to 45 degrees. All of this functionally comes standard with any table saw. The possibilities multiply when you incorporate jigs and fixtures: box-joint jigs, tapering jigs, cove-moulding jigs, raised-panel jigs, tenoning jigs, crosscut sleds and more. The possibilities are only limited by your imagination, and perhaps by your engineering know-how.

To match the versatility of the table saw in the hand-tool world, you would not only need numerous tools (large and small saws, large and small handplanes, chisels and jigs), but also the time, patience and skill to use them effectively. If you choose to take the more difficult path of all hand tools, I applaud your dedication. But for the average woodworker it's going to be hard to argue with the easy learning curve and time savings associated with the most basic of table saws. For that reason, I think it's safe to say that the table saw is still the heart of the typical woodshop, whether it is all-power or hybrid.

The table saw does have a couple of disadvantages that could matter to some hybrid woodworkers: It takes up a lot of space in the middle of the workshop and it can be dangerous, especially to the inexperienced and occasional user.

Tenoning Jig ▪ A commercial tenoning jig holds the workpiece vertically for the cheek cuts, yielding smooth and consistent results. It's a great jig for batching out a large number of tenons.

Depending on the mix of projects in the shop, the space issue can be addressed with a sliding-arm miter saw mounted on a bench along one wall, or with a portable jobsite saw that folds or rolls out of the way when it's not in use. And whatever table saw you own or plan to purchase, be on the lookout for high-quality safety gear. A splitter is good but a riving knife is better. A riving knife not only keeps wood from migrating into the back end of the blade and kicking back dangerously, it also travels up and down with the blade so there are very few times you'll need to remove it from the saw. Blade guards are also important as a brute-force protection method: If your fingers can't contact the blade, you won't get cut. But keep in mind, inexpensive saws usually have clunky blade guards. From my experience, trying to use the saw with these guards in place can feel more hazardous than doing without. So inspect the guard carefully and make sure it is well-built, reliable and easy to remove and replace when necessary.

The Table Saw in My Workshop

My first table saw was a Craftsman contractor model. It was a gift from my wife and a not-so-subtle hint for me to get moving on her honey-do list. Although underpowered, the saw performed well after I outfitted it with a thin-kerf blade that eased the strain on the motor. Over time, as my woodworking skills and overall awareness of quality improved, I noticed that the fence wasn't consistently locking down parallel to the blade. In fact, I had been blaming my poor results on what appeared to be a lack of skill and I didn't realize the tool itself might be to blame. I come from the school of thought that believes in the old saying, "A poor carpenter blames his tools." I will always look inward before I blame any outside influence. But if the tool is truly flawed, the old saying doesn't apply. I tried everything I could to improve the operation of the fence and nothing worked. I wasn't in a position to buy a new saw, so I scraped up the funds for an after-market fence

Thin Is In ■ Thin-kerf blades save material and motor strain.

Thin-Kerf Blades

If your saw has a motor rated less than 3hp, you can get a serious performance boost by using a thin-kerf blade. The teeth on a standard full-kerf blade are about 1⁄8" wide. If you think about it, that's a lot of carbide to push through a dense hardwood. Thin-kerf blades typically measure about 25 percent narrower, about 3⁄32". This reduction in size makes it much easier for the blade to plow through thick stock even if the saw isn't a 5hp beast with a flame paint job. Less resistance means less friction, less strain on the motor, less burning of the wood and overall better results. So if your saw feels like it's bogging down on thicker cuts, switching to a thin-kerf blade can make all the difference.

One word of warning: If you decide to switch to a thin-kerf blade, it is likely you'll need a new splitter or riving knife. A splitter or riving knife intended for a full kerf is likely to get stuck in the narrower slot made by a thin-kerf blade.

system. The new fence worked quite well and the quality of my results improved dramatically. As my hobby evolved into an obsession and then later into a career path, I upgraded the saw to a full-size 5hp Powermatic cabinet saw (PM2000).

I hesitate to recommend a saw because I know many folks who do amazing work with compact jobsite table saws that others would consider cheap and underpowered. But personally, my shop life improved dramatically the day I received my full-size cabinet saw. For a professional shop, it's a smart investment. But for hobby shops, it's more a matter of budget and personal preferences. The table saw enables all the various cuts required by the hybrid woodworking method. Keep it calibrated and keep it clean, because you'll be using it a lot.

A Router in Every Shop ■ The router is easily the most versatile hand-held power tool in the shop. Whether it's joinery, milling, edge treatments or embellishments, the plunge router can do it all.

Router

The router is to the portable power-tool world what the table saw is to the stationary power-tool world: a multi-tasking powerhouse. It's ironic that a tool capable of creating such a dizzying array of profiles and joints is so simple in design: a cutting bit mounted in a small motor that rotates it. Milling, joining, treating edges and setting decorative inlays are just a few of the things a router can do. Should you decide to take things to the next level and mount the router upside-down in a table with a fence, you'd have the functionality of the wood shaper as well.

How I Use the Router

When I purchased my first router, I used it for nothing more than edge treatments such as roundovers and chamfers. It wasn't long before I realized the tool's potential in the world of joinery, and suddenly I saw the router in an entirely different light. With the help of an edge guide (included with most routers), it's a piece

Poor Man's CNC? ■ While not quite a CNC machine, with the right setup a router can flatten large and unwieldy boards that won't fit through the thickness planer.

Router Table ■ Mounted under a router table, the router becomes even more versatile for shaping, joinery and edge treatments.

A Compact Option ■ Small palm routers are handy for adding a final round-over to your projects or when working in tight spaces.

of cake to rout clean and consistent mortises quickly and efficiently. The router also excels at cutting dados, rabbets and grooves, especially in situations that aren't feasible or safe at the table saw. Whenever I have a large number of drawers to make, I rely on the router and a dovetail jig to get the job done. If an extra-wide board requires flattening, I use a router jig on parallel rails to flatten the surface, much like a manual CNC machine. This is how I flattened the 96" x 24" top of my workbench and it worked like a charm. On the more delicate side of things, the router can also be used to make elegant and artistic decorative inlays.

Routers in My Workshop

Woodworkers are very good at turning wants into needs. In reality, a woodworker needs only one router. A good quality 2¼ hp plunge router will do just about every job you ask of it. But because of the wide range of things a router is typically used for, we can certainly benefit from having a few routers, each dedicated to a particular task. This is why you'll often meet woodworkers with 10 or more routers in their collections. While I feel 10 or more is overkill, I do think most woodworkers would benefit from having three: a plunge router, a dedicated heavy-duty router for the router table and a small palm router for light-duty detailing.

A 2¼ hp variable-speed plunge router that can take both ½" and ¼" bits is perfect for any profiles or joints that need to be cut by bringing the tool to the wood (as opposed to bringing the wood to the tool). For the router table, I like to have a more powerful dedicated machine of at least 3hp to handle the larger range of bits including big panel-raising and pattern bits. The third router I recommend is on the opposite side of the spectrum from the beefy table router: a little palm router. Some routing operations need to be done in tight spaces and the palm router is the only tool that will fit. Whenever I have a stack of project parts that needs a slight edge treatment, the palm router is the perfect tool for the job because it is lightweight and can be used with one hand, resulting in much less fatigue.

Band Saw

The band saw is an essential woodshop cutting machine that can saw both straight lines and curves. The saw itself is simple, consisting of a long continuous blade that rides on two large wheels. The work surface is positioned between the two wheels and supports the workpiece as it's pushed through the blade. Bandsaw blades come in numerous tooth configurations and sizes for a wide variety of tasks and materials.

It should be noted that with growing concerns about table saw safety, some woodworkers, including hybrid woodworkers, are ditching their table saws and instead relying heavily on the band saw. This requires some creative solutions and alternative tooling but the safety trade-off is certainly worth it for some folks. I won't be selling my table saw any time soon but I do find myself using the band saw more and more. That says a lot for this tool's range of functionality.

How I Use the Band Saw

I use the band saw for four tasks: ripping, re-sawing, cutting curves and cutting joints.

Ripping is sawing a length of wood into two or more narrower lengths. It's traditionally the job of the table saw, but to be a safe operation, the wood must have at least one flat face and one flat edge, otherwise a dangerous kickback could occur. Additionally, wood tends to bend after a long rip cut and the chances of kickback are increased if the wood bends into itself, pinching the back of the table saw blade. So while the wood is still in a rough state, I like to do the initial rip at the band saw. If the wood warps after the cut, it won't cause kickback and I have an early warning sign that this particular board might be temperamental. Only after the wood

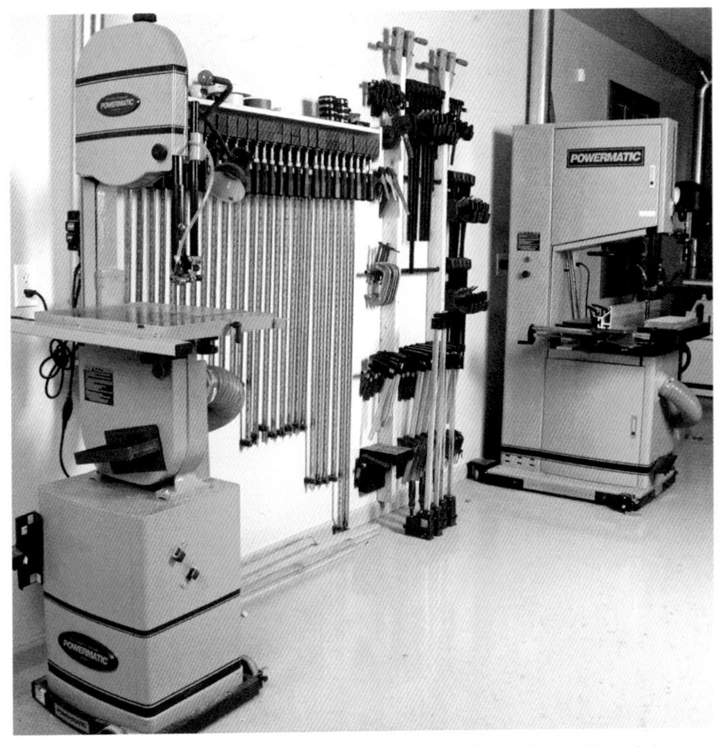

Double Your Fun ▪ My shop has two band saws: a 14" saw for cutting tight curves and a 24" machine for resawing and wide curves.

Resawing ▪ With a 3hp motor and a sharp blade, the 24" band saw makes quick work of slicing thin veneers from wide boards.

has been milled flat and square will I use the table saw to trim the board to final width.

Resawing is just like a rip cut only the goal is to saw a single board into two thinner boards. Using this technique, we can actually make our own veneer by slicing off super-thin sheets of wood one at a time. Unlike commercial veneer, which is usually cut to 1⁄32" thick, shop-made veneer usually comes in about 3⁄32" thick after sanding, which provides a much more durable product when it's properly glued to a stable substrate. Incorporating veneer into your work will open up a whole new world of possibilities in materials and design and helps conserve highly figured and rare wood species. It's a great feeling when you can take one single prized board and make an entire tabletop out of it.

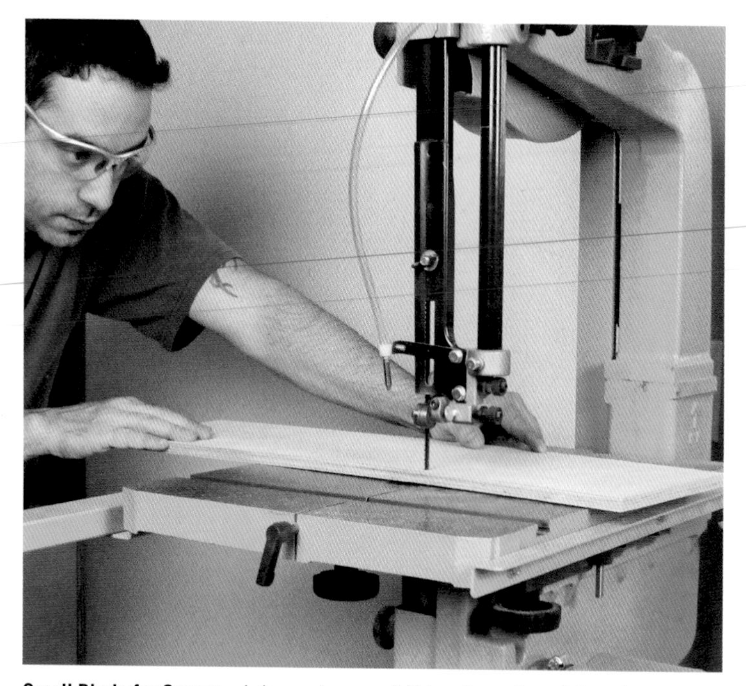

Small Blade for Curves ▪ I depend on my 14" band saw for cutting the various curves featured in my work. A narrow blade can make tight turns.

Cleanup Cuts ▪ The 14" band saw excels for various clean-up cuts, such as removing the waste between bandsawn dovetails.

Any time a workpiece needs a curve cut, the band saw is the go-to tool. With a narrow blade, the band saw can be used like a scrollsaw, making tight turns for complex patterns. In my shop, most curves I cut are fairly wide and long and even a 3⁄4"-wide blade will usually get the job done.

Cutting joints at the band saw can be a lot of fun but sometimes requires thinking outside the box. Two popular woodworking joints that come to mind are dovetails (see techniques section) and tenons. Using hybrid methodology on the band saw, both joints can be cut quickly and effectively with a quality that rivals any other method out there.

Band Saws in My Workshop

I currently own two band saws: a 24" model and a 14" model. The 24" saw is dedicated to ripping and resawing and is outfitted with a 3⁄4"-wide carbide-tipped blade. The table is beefy and wide and can handle the abuse of large, rough timbers. I use the 14" saw for cutting curves and other general sawing tasks. It's outfitted with a 1⁄4"-wide blade to accommodate tight curves. Most woodworkers can easily get by with one band saw but it might be necessary to change out the blade for some operations. If I had only one saw, thanks in part to a hefty dose of laziness, I would outfit it with a good general-purpose blade with the following specs: 1⁄2" wide, 3 teeth per inch, hook configuration, minimal set. A blade like this is narrow enough to handle most curved work, has lots of big gullets for dust to escape the cut (which helps reduce friction and motor strain) and minimal set to the teeth, which creates a narrow kerf and a cleaner cut.

Save Your Back ▪ Break down sheet goods on the shop floor with the workpiece resting on foam insulation sheets. A portable circular saw and shop-made guide makes break-down quick and easy.

The Track Saw Option ▪ If you cut a lot of sheet materials, consider a track saw. This Festool yields table-saw quality cuts with exceptional accuracy and effective dust collection.

Portable Circular Saw, Track Saw

The concept of the portable circular saw is scary when you think about it: a blade spinning at 5,000 rpm just a few inches from your hand. In many ways, it's like a hand-held table saw. Fortunately, when used properly, the tool is perfectly safe and incredibly handy. A newer variant of the circular saw is the much-adored, but pricey, track saw. Track saws are made to ride securely on an aluminum track for perfectly straight table-saw quality cuts. With an upgraded blade and a homemade saw guide, a standard circular saw can be used in a very similar fashion.

How I Use Portable Circular Saws

Whether using a portable circular saw with a guide or a track saw, the main job of this tool in my shop is to break down sheets of plywood and MDF. As a chronic sufferer from lower back pain, I have absolutely no desire to lift 4x8 sheets of plywood onto the table saw. Instead, I lay the sheets on the floor over a few pieces of rigid foam insulation. I then use the portable circular saw to rough-cut the sheet into smaller, more manageable pieces, which can then be trimmed to final size at the table saw. Once you try this break-down method, I guarantee you'll never try to wrangle a large sheet of plywood onto the table saw ever again.

Additional uses for a portable circular saw include breaking down planks of solid wood (something I prefer to use the jigsaw for), as well as unique one-off cuts. A good example of a one-off cut was when I had to trim my 4"-thick workbench top to length. The slab was far too large for the table saw or the miter saw, so my only choice was to make the cut in two passes using the circular saw. The first cut went about halfway through the thickness of the top, then I flipped the slab and made a second cut. The resulting surface wasn't perfect and needed a little block-plane love, but it was certainly good enough for the end of a workbench. For unique jobs like that, the portable circular saw is very nice to have around.

Portable Saws in My Workshop

I own both a regular portable circular saw and a track saw, specifically the Festool TS55. These two tools are redundant because the track saw can do anything the circular saw can do, and more. Because I often demonstrate alternative tools and methods in my instructional videos,

Jigsaw for Curves ▪ A jigsaw is adept at cutting curves. It's a great substitute for a band saw if you can't justify the cost or space.

Jigsaw for Rough Cuts ▪ When breaking down large rough boards, I find the jigsaw simple, safe and predictable.

it's important for me to keep the circular saw around. As great as track saws are, they can be expensive and not everyone can justify the purchase. Fortunately, as the popularity of the track saw grows, more companies are getting into the game. As a result, prices are coming down and more budget-friendly options are hitting the market. If you already have a good portable circular saw, an upgraded blade and a shop-made guide are all you truly need to add.

Jigsaw

The jigsaw is a small portable saw that excels at cutting curves. When assembling a new shop on a budget, I often recommend that people buy a good jigsaw before investing in a full-size band saw. At least initially, most folks use a band saw for curve cuts and that is something the jigsaw does quite well while taking up much less space and costing significantly less money. Outfitted with the proper blade, a jigsaw can cut just about anything from metal to plastic to any sheet material you can think of, except maybe granite. Unfortunately, the cut quality isn't always great and the surface will usually require subsequent finessing. But that's why we have a nice assortment of hand tools, isn't it? Even after you add a band saw to the shop, a jigsaw will continue to come in handy for various one-off cuts and the break-down of rough stock.

How I Use the Jigsaw

Because I already own a band saw, I don't often use the jigsaw for anything other than breaking down rough lumber. People are often surprised when they see me do this but I have a very good reason for it. Cutting rough lumber can be scary with a spinning blade. Circular saws, miter saws and table saws can all kick back if the lumber moves during the cut. The rougher the board, the more likely it is to shift or pinch the blade. The jigsaw, which has a simple up/down blade motion, removes all possibility of kickback. The tool is light and easy to handle, and makes breaking down boards fast and easy. Because the wood is in a rough state, I like to use an aggressive blade that cuts quickly and efficiently. Once I've sawn the boards into manageable sizes, I can proceed with jointing, planing and the rest of the milling process.

Jigsaws in My Workshop

You don't need a fancy jigsaw for roughcutting lumber. Its primary job is to hold a blade securely and even the cheapest jigsaw on the market can typically handle that. But if you think you might need to do more curve-cutting on critical workpieces, or if you plan to use it as a substitute for a band saw, you should invest in a decent contractor-grade model.

Jointer and Thickness Planer

Many woodworkers begin their journey into the craft by buying pre-milled boards from the home center. These boards are already milled to thickness and supposedly sport flat and parallel faces and edges. Unfortunately, even if the boards were perfect at the mill, they probably aren't after sitting on the store rack for several weeks or months. It isn't long before new woodworkers realize the benefits of buying boards in the rough from a local hardwood dealer (which saves money), and milling it themselves (which saves frustration).

The two primary milling tools are the jointer and the thickness planer. They can be purchased individually or together in a combination machine. People often ask me which one they should buy first. Given the price of these tools, it's quite common to only invest in one at a time. Without a doubt, the best initial bang for the buck is the thickness planer. Most rough boards from the mill are still pretty close to flat, so you can often get away with sending them through the planer, one side at a time, making very light cuts to bring the board to the desired thickness. This method is called skip planing and can yield decent results if the starting material is in pretty good shape to begin with. However, if the board is warped or twisted, the thickness planer will simply follow those flaws instead of correcting them. If you choose to go the planer-only route, this is an important factor to consider when purchasing your rough stock.

How I Use the Jointer and Thickness Planer

The thickness planer addresses the faces of the board, but what about the edges? There are numerous alternative methods for getting a straight and square edge, including jointing with a router, trimming with a sled on the table saw and truing the first edge the good old-fashioned way, with a jointer plane. In fact, these methods

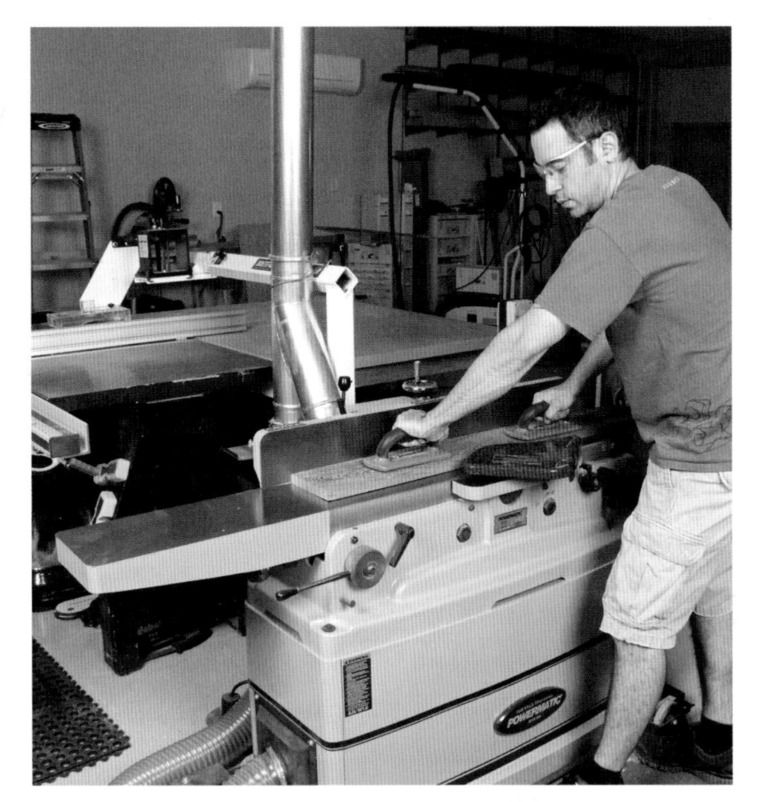

Jointing the Face ■ You won't get far in woodworking without flat and square boards. The jointer is the go-to tool for creating a flat face on a rough board.

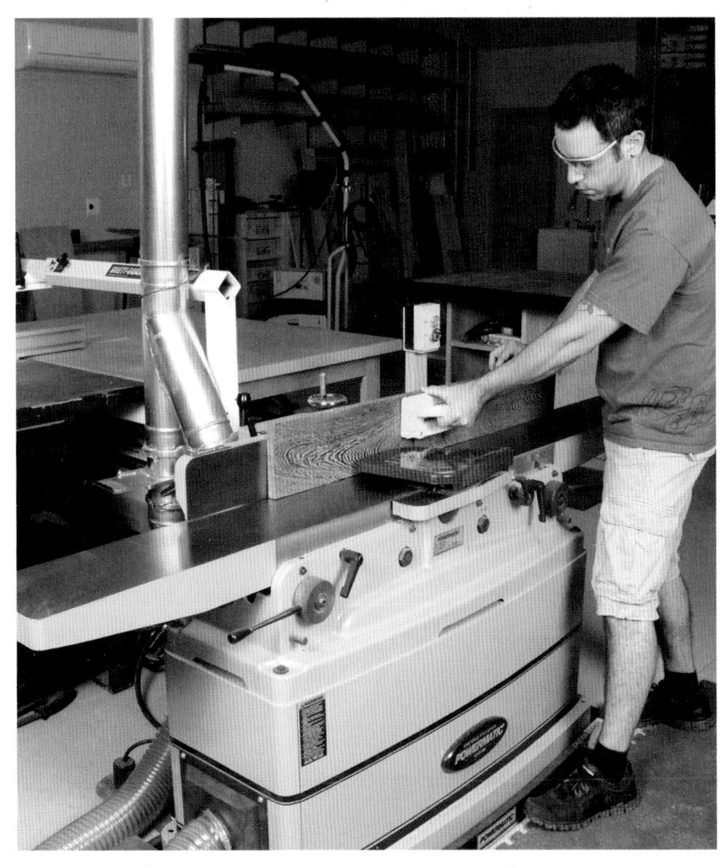

Jointing the Edge ■ Press the flattened face against the fence to joint one edge of the board straight and square.

can be so reliable that some woodworkers thrive with a thickness planer alone and never feel the need to purchase a jointer.

While I certainly could survive without my jointer if I had to, I really don't want to. The jointer serves the vital purpose of establishing flat reference surfaces that ultimately result in truer, flatter boards. The jointing process usually begins by making one face nice and straight and flat. With the flat face against the fence, the machine will plane one long edge straight and square to the face. The woodworker can then send the board through the thickness planer with the flat side down, providing the perfect reference for the planer to create a parallel and flat face on the other side of the board. This produces better and more reliable results than the skip-planing method. At the table saw, the board can then be cut to width using the jointed edge as reference against the fence.

While some people can get away without one, I personally find the jointer to be absolutely essential for turning rough wood into flat and square project boards.

Thickness Planing ▪ The thickness planer makes quick work of bringing a board to the desired thickness.

Skip Planing ▪ Skip planing can get wide boards mostly flat by taking light passes on each side.

Jointer and Planer in My Workshop

I use a Powermatic 8" jointer. I often receive emails from woodworkers wondering whether they should buy a 6" or an 8" model. For me, the answer is clear: 8". Most of the wood I purchase is in between 6" and 8" wide. If my jointer were only 6" wide, most of my wood would be too wide and would require extra work to mill flat and square. But an 8" jointer, with its increased capacity, can handle about 90 percent of what comes into the shop. As I see it, an 8" jointer hits the sweet spot. There are wider jointers on the market, but they get significantly more expensive, making the 8" jointer the best bang for the buck.

My thickness planer is a 15" Powermatic model. You might be interested to know that I used to own a similar 20" planer. As it turned out, I hardly ever used the full capacity. To understand why, let's think about the logic for a moment. I only have 8" of capacity with my jointer, so the widest board coming off of the jointer is going to be 8". Of course, there may be times when laminated boards or extra-wide boards need to be sent through the planer, but those times are few and far between. So I sold the 20" planer and downgraded to a 15" model. I have yet to confront a situation where I wished I had more capacity and the 15" planer has a much smaller footprint. Bigger isn't always better.

Miter Saw, Sliding Compound Miter Saw

The compound miter saw is the ultimate crosscutting tool. Whether the cuts are 90 degrees, mitered, beveled or a combination of the two, as you'll find in things like crown moulding, the miter saw can do it all. A variation of the miter saw is the sliding compound miter saw where the blade is capable of traveling forward on a set of rails or with an articulated arm. This feature increases the capacity of the saw, making it even more flexible for crosscutting parts. With a good carbide blade installed, it is possible to achieve near finish-quality cuts using this tool.

How I Use the Miter Saw

The table saw also can make crosscuts, so owning both machines is redundant. But for any workpiece that fits within the tool's capacity, I favor the miter saw because it is so quick, easy and accurate. With extension fences and stops, the tool can batch-cut multiple parts to an exact length. Only when the width of the workpiece exceeds the capacity of the miter saw do I change to the table saw instead.

Miter Saws in My Workshop

I have owned a few different miter saws over the years and they all performed quite well. If the saw cuts accurately and cleanly at 45 and 90 degrees, both miters and bevels, that's about all I require. Most budget-friendly saws will perform adequately with nothing more than a calibration and a blade upgrade. To improve the results, consider installing an auxiliary fence and a zero-clearance insert. Just like at the table saw, a zero-clearance insert will support the workpiece and help minimize tear-out.

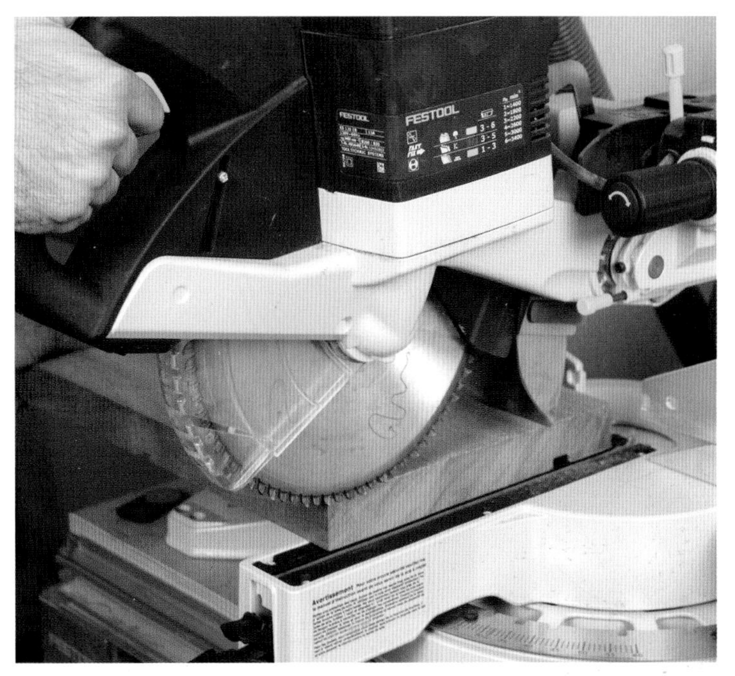

Sliding Compound Miter Saw ■ A miter saw is great but a sliding compound miter saw is better. Its increased capacity can crosscut wider boards.

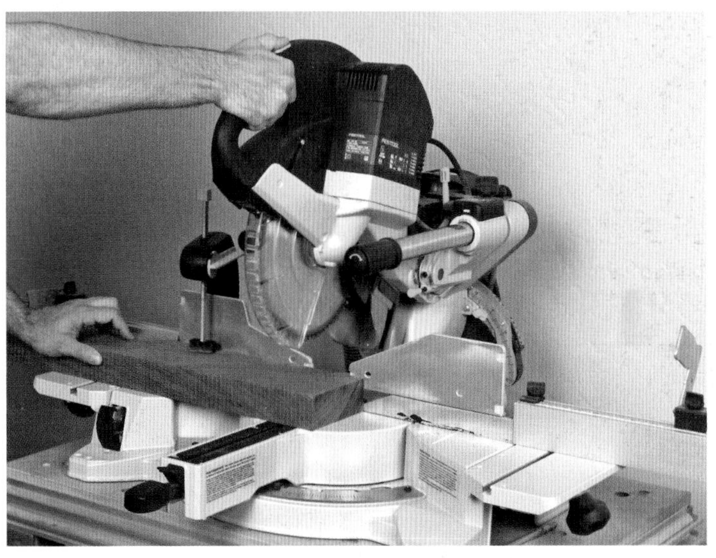

Miter Saw Add-ons ■ With fence extensions, stops and clamps, the miter saw can be used to make repeatable crosscuts and accurate miters.

Random-Orbit Sander

The random-orbit sander is a popular sanding tool that not only vibrates but also oscillates in a randomized pattern, creating scratch marks that are hard for the eye to detect. In the past, I would use the random-orbit sander for all of my surface preparation, from rough sanding to final smoothing. But now I typically use it for the final #220-grit sanding only, thanks to the help of my hand tool friends: scrapers and planes. Minimizing machine sanding is one huge benefit of hybrid woodworking.

Regardless of how much sanding you choose to do, it's important to know how much sanding is enough. I can't tell you how many times I have been asked, "How do I know when I'm finished

Random-Orbit Sander ■ Sanding is a necessary evil. A random orbit sander with good dust collection makes the task tolerable.

with a particular grit?" Whenever possible, I like to employ a systematic solution that any woodworker can follow, and that's the type of solution I use to gauge sanding progress. It's so simple, in fact, that it requires nothing more than some pencil marks on the surface.

How I Use the Random-Orbit Sander

Before sanding I use a blunt pencil to lightly scribble a series of pencil marks across the face of the workpiece. The pencil dust not only sits on the surface but also settles into some of the scratches and imperfections left by the tools and previous sanding grits. I begin sanding slowly and evenly, moving at a rate of about 1" per second, and I don't stop with that particular grit until it has completely removed the pencil. I then swap the paper out for the next-higher grit and once again mark the surface with the pencil. Sanding resumes and doesn't stop again until all the pencil is gone. With each higher-grit sanding, the surface scratches get smaller and the pencil dust gets removed more quickly. Without a system like this, it is very difficult to ensure the surface is thoroughly and evenly sanded and ready for the next grit in the sequence.

Random-Orbit Sanders in My Workshop

I own several random-orbit sanders but the one I use the most is a 6" model. When it comes to sanding, I want to cover as much real estate as possible with no more effort than necessary, so the bigger the pad, the sooner I can move on to the next phase of the project. If possible, try to find a sander that accepts multiple types of sanding pads. When sanding smaller surfaces, it can be handy to have a rigid pad that won't round over the crisp edges. And because sawdust is a major health hazard, always use a shop vac or dedicated dust extractor with the sander. Most models have a dust port that can be attached to a vacuum hose. I never run a sander without dust collection, ever.

Breathe Easy

Tool ownership brings with it a certain set of responsibilities. Norm Abram of *The New Yankee Workshop* said it best at the top of every show: "Be sure to read, understand and follow all of the safety rules that come with your power tools." And although most of us know to be careful around spinning blades, there's another major safety concern that is all too often overlooked: dust collection. Each and every one of my power tools is hooked up to a network of rigid ductwork that leads back to a 5hp Clear Vue cyclone. Air cleaners can be helpful and shop vacs are great for cleaning up a mess, but nothing beats collecting the dust at the source. If it never hits the air in the first place, the shop will stay cleaner and there will be fewer fine dust particles to breathe.

The main ductwork runs down the left side of the shop ceiling and branches off to each set of tools. There's a blast gate at or near each tool to help isolate and control suction. Any tools that aren't connected to the cyclone system are connected to individual dust extractors, such as the miter saw, the routers, sanders and portable saws.

In spite of my rather capable dust collection setup, I still wear my 3M 7500 respirator when using the big dust producers. The respirator features a set of goofy pink particulate filters that I consider a badge of honor. It says to the world, "I care about my health so much that I'm willing to wear a pink respirator!" Seriously though, no matter how good your dust collection is, fine dust will always escape into the shop air. So to be on the safe side, wear additional protection.

I'm 36 years old and I fully expect to be working with wood for the rest of my life. So protecting my lungs is critical.

Continued on next page.

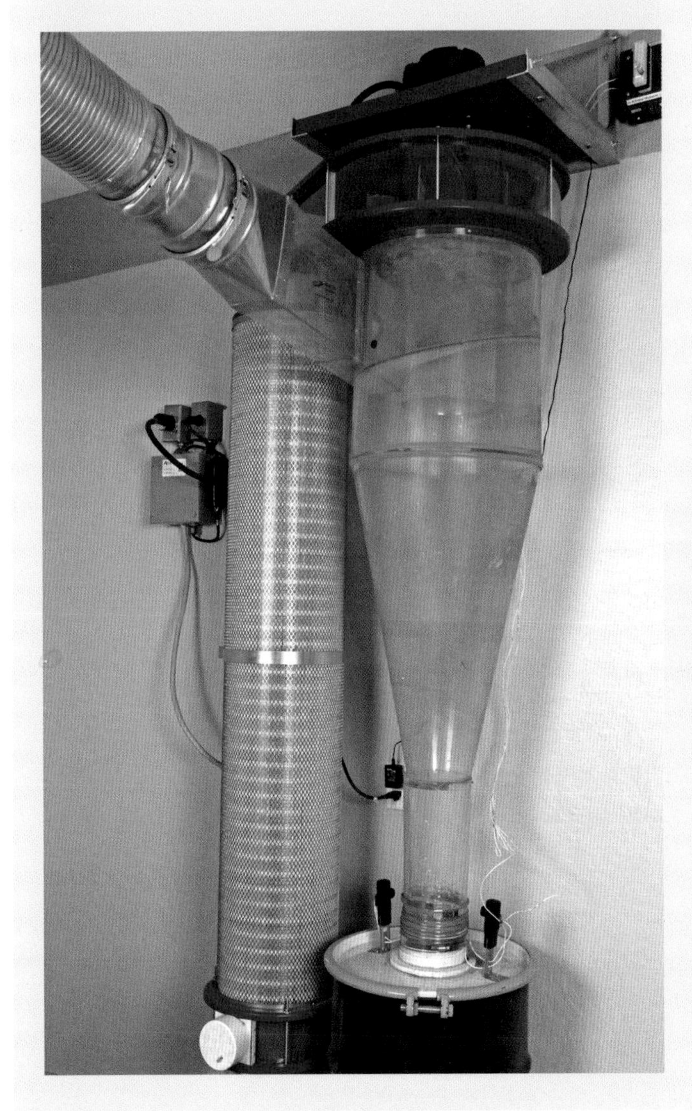

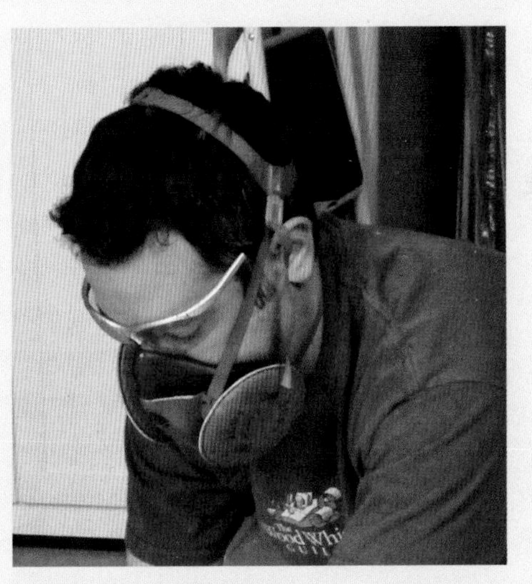

Respirators – Fail-Safe Option ■ Even with good dust collection, I wear a respirator to protect me from the smallest and most harmful dust particles.

Cyclones Have the Power ■ If you plan to run a ducted dust-collection system, you should consider investing in a cyclone for the ultimate in power and efficiency.

Breathe Easy

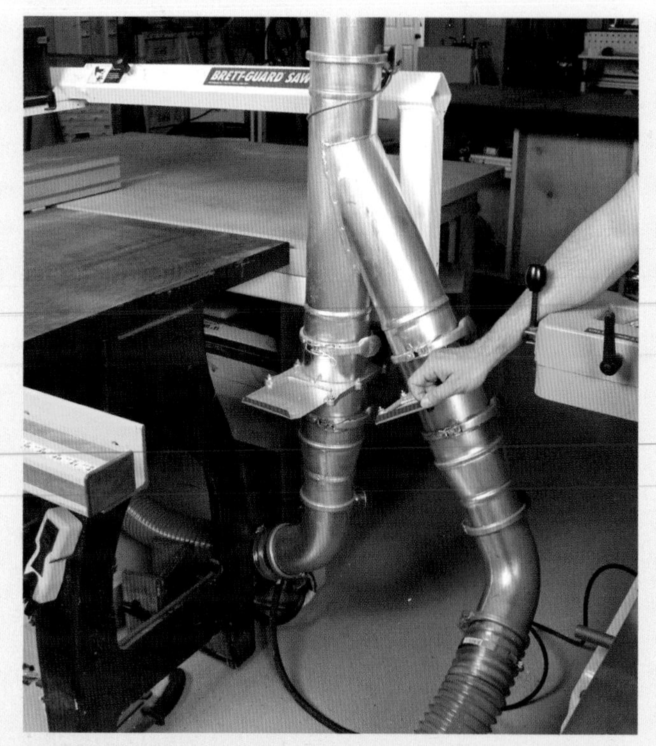

Blast Gates Increase Airflow ▪ Each tool has its own blast gate, which remains closed when not in use. This redirects the suction to the tools that need it.

A Big Dust-Maker ▪ When resawing, the band saw creates an incredible amount of dust. Collecting the dust at the source keeps it from getting into the shop air.

Portable Dust Extraction ▪ When possible, connect portable power tools such as the router and sanders to a dust extraction system or a shop vac.

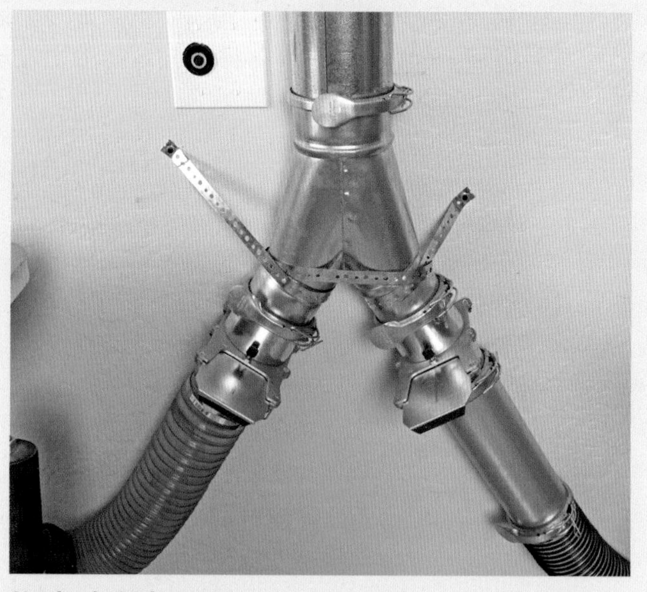

Sharing is Caring ▪ Many of my machines are clustered together so I use Ys to branch two lines off a single drop.

Other Power Tools

There are a number of power tools that I rely on that don't necessarily relate directly to the hybrid process, but I thought they were worth mentioning as possible additions to your woodworking arsenal.

Drill Press

Most woodworkers own at least one power drill, if not multiples. And while you can do quite a bit with a hand-held drill, there's nothing like the versatility and accuracy of a drill press for fine woodworking. There are many times when drilling a perpendicular hole is absolutely essential to a project's success, and that's where the drill press excels. With the aid of a fence and a depth stop, you can drill perfectly perpendicular repeated holes, quickly and easily. Also, some specialty bits are much safer to use at the drill press, such as Forstner bits, which walk along the workpiece when used in a hand-held drill.

Drill Press ▪ Hand drills are great but a drill press offers accuracy, repeatability and versatility.

Oscillating Spindle Sander

The oscillating spindle sander is incredibly helpful to the curve-making woodworker. Any time you need to smooth a curve, the spindle sander can make quick work of the job. With several spindle diameters, you can match the radius of the curve that needs sanding, or simply use the largest one to finesse wide curves. While you can finesse curves with hand tools at the workbench, the oscillating spindle sander can do the job in much less time.

For the Curve-Maker ▪ My work has a lot of curves and the oscillating spindle sander helps me do the initial smoothing.

Biscuit Joiner

A biscuit joiner is a handy tool to have around. While I don't recommend biscuits for structural purposes, they do work well as alignment aids. For example, when gluing two boards together to make a tabletop, the long-grain glue bond alone is plenty strong. But getting those boards aligned perfectly during the glue-up can be a challenge. The bigger the table and the more boards, the trickier this process becomes. Biscuits offer a fantastic solution. By creating narrow crescent-shaped mortises in

Alignment Made Easy ▪ A biscuit joiner makes for dummy-proof glue-ups by keeping the work aligned.

each adjoining edge, a biscuit inserted between the boards acts as a small spline, keeping the pieces from moving. Truth be told, I don't use my biscuit joiner much these days, thanks to the Festool Domino. The Domino can make high-quality loose-tenon joints, and the smallest Dominos are perfect for these alignment tasks.

Hollow Chisel Mortiser

The hollow-chisel mortiser does one job and it does it well. The way the tool works is very similar to a drill press as it features a spinning bit that you manually plunge into the work. The mortiser has one very important addition: a square chisel that surrounds the bit. So as the bit hogs out the bulk of the material, the chisel is right behind it chiseling the hole square. Advantages include fences and stop-blocks for batching out project parts and the fact that square mortises require no rounding over of tenons to fit. There are a couple of disadvantages you should be aware of as well. First, the tool is a uni-tasker. If you have limited shop space, you might not want to devote valuable floor or countertop real estate for a tool that only does one single operation. Also, the tool is a little finicky to set up and calibrate. The chisel and bit must be set precisely and the chisel needs to be perfectly square to the fence. Once you get the hang of it, setup does become easier. If you make a lot of mortises in your work, this tool is worth your consideration.

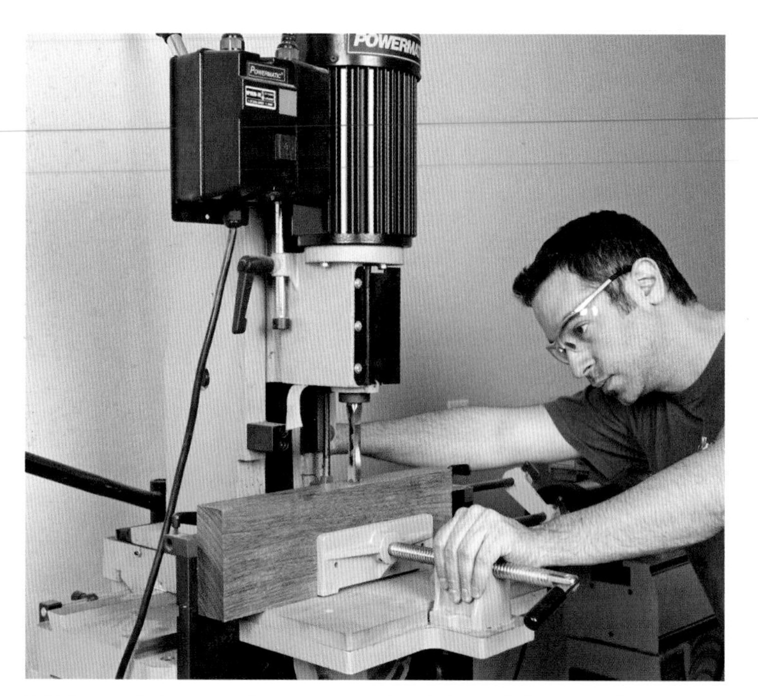

Drill Square Holes ▪ A hollow-chisel mortiser is like a drill press with a square chisel surrounding the bit. It can cut square mortises with repeatability.

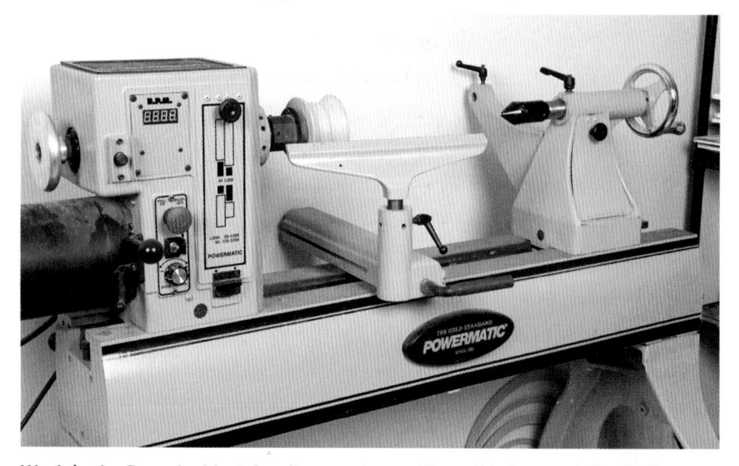

Work in the Round ▪ Most furniture makers will want to have a lathe in the shop for turning furniture parts and components, as well as for making beautiful bowls, vessels and pens.

Lathe

Let me start out by admitting something: I don't use my lathe all that often. I have friends who love spending entire days in the shop turning bowls, pens and other beautiful works in the round. And while I do enjoy turning something on occasion, I find myself much more interested in furniture construction and joinery. But that doesn't mean I don't need a lathe. There are just some furniture parts that can't be made any other way, at least not via practical means. For example, some designs call for turned spindles, legs, knobs and other elements crucial to the chosen design. Without a lathe, I'd have to have someone else make these parts for me, and that's no fun. So regardless of whether you call yourself a turner, a lathe is a good investment for any furniture maker.

Woodworking NInja ▪ My hand tools make me feel like a woodworking ninja. I am one with the steel.

Must-Have Hand Tools

Hand tools are nothing new, but to the modern woodworker their uses and benefits may become obvious only after we've accumulated a whole shop full of power tools. So to many a new woodworker, hand tools are the latest and greatest thing since sliced veneer. This is not a phenomenon isolated to new people entering the craft, either. Many folks who've been calling themselves woodworkers for years are just now discovering the joys of sharp chisels and gossamer plane shavings. Thanks in large part to the Internet, many outspoken hand-tool proponents are sharing their woodworking methods and viewpoints with the entire world, and as a result we are experiencing a hand-tool renaissance.

One doesn't need to look too hard to find evidence of this recent trend. Woodworking magazines have begun to present a more balanced focus featuring both hand- and power-tool methods and reviews. Countless woodworking blogs are dedicated solely to the pursuit of hand-tool greatness. The hand-tool sections of most popular woodworking forums are busier than they've ever been. Most recently, several notable woodworkers have begun offering paid online courses with a strong focus on hand-tool woodworking.

Numerous small manufacturers are riding this wave of renewed interest in hand tools and the choices can be mind-boggling. Lie-Nielsen Toolworks is well known for taking old, retired hand-tool designs and giving them a new lease on life. Lee Valley Tools also like to bring back old designs, except they add a modern spin that many woodworkers appreciate. Super-premium boutique dealers such as Bridge City Tool Works, Sauer & Steiner Toolworks, and Brese Planes make exquisite tools that can only be described as art and that cost more than my monthly mortgage payment. Even small independent woodworkers are making a go of it, building their own brand of wooden-bodied handplanes, like my friend Scott Meek of Scott Meek Woodworks.

How about our favorite woodworking events? Just a few years ago, every woodworking show I attended focused almost exclusively on power tools. Today, one of the most exciting and popular events is the yearly Woodworking in America conference, which began as a hand-tool-only event. Although Woodworking in America now has its fair share of power tools, it still boasts a heavy dose of hand-tool bias in both its seminars and vendor selection.

It has never been easier to learn about and acquire hand tools. In fact, there is a distinct and somewhat unfortunate possibility that after doing some homework, you'll find yourself in perpetual research mode, suffering from paralysis by analysis. To make matters worse, once you decide on the type of tool you need, you have to think about how much money you want to spend because there's a brand to match every budget. Fortunately, the tool-buying guidelines I provided previously should help you through just about any purchasing quandary. But before you get to that point, you should probably have an idea of what a good, functional hybrid woodworker's tool kit looks like. The tools listed below are the ones I find myself relying on for every single project. Throughout my years as a woodworker, some tools moved out while others moved in. Some tools share functionality but approach it in different ways, so you might notice some overlap. Remember, it's all an evolution, and it's very personal. While this may not be the end-all-be-all tool list, it should serve as a starting point for strategizing your own collection of hand tools for hybrid woodworking.

Smooth Moves ▪ Both of these are smoothing planes, but the hand-made wood-bodied plane on the left feels very different in use than the low-angle metal-bodied smoother on the right.

Meeting of the Minds ▪ Woodworking in America is a yearly educational conference that celebrates both hand and power tools.

Sharpening

Sharpening is one of those topics where if you ask 20 woodworkers what they do, you'll get 20 different answers. To make matters worse, all twenty methods probably work quite well. Don't let the array of options bog you down. Instead, find a method that works for you and stick with it. Early in my career, in search of the best option, I made the mistake of trying too many different systems at once. The truth is, there is no best option. Some are certainly better, easier, faster or less expensive, but nothing is best for everyone and every situation. Your primary goal should be finding a system that gets you from dull to sharp with as little hassle as possible. If

What Tool to Buy First?

It's an age-old question uttered hundreds of times a year in my inbox: What hand tool should I buy first? Without context, the question is impossible to answer. Most times, I fire back several questions that help people understand why I can't give them specific recommendations yet. What tools do you already own? What do you want to build? Do you want to mill lumber and will you use hand or power tools to do it? The trouble is, most new woodworkers don't know the answers to these questions, and therein lies the difficulty. Tool-buying decisions depend on what the user plans to do on his or her woodworking journey, which is something probably unknown at the outset. Because you are reading this book, I'm going to assume you are interested in the hybrid method and you already have a decent complement of power tools. Armed with that information, here are a few recommendations:

Stay Sharp ▪ You need sharp tools, so find a good sharpening system and stick with it.

A Sharpening System

Yes, I realize that it's anticlimactic to recommend sharpening equipment as a first purchase. It's like telling new teenage drivers that before they can have the keys, they have to learn how to change the oil. But just like a car, hand tools can't function without regular maintenance in the form of properly sharpened blades. If you don't have a rudimentary understanding of sharpening, your hand tools will be nothing more than cool-looking paperweights. For years, I avoided hand tools primarily because I never had good results. My block plane seemed to have a ferocious set of teeth that would bite aggressively into the wood surface. The heart of the problem was with the tool, but the fault was all mine. If I had taken the time to learn how to sharpen the blade, the tool would have performed well enough to encourage me to use it more, which would have driven me deeper into the world of hand tools. This is why I think acquiring sharpening skills is the first step in anyone's journey into the hand-tool world.

Chisels ▪ Chisels are essential for cutting good joints.

Chisels

A set of chisels is an absolute requirement for woodworking. Most woodworkers already own a set, even if they only use them as can openers. Chisels are assets for not only making joints, but also for general multitasking. There are many times during a project when something needs trimming or shaping and the chisel is the only tool that can do the job. Once you know how to wield them, chisels will become your solution to a lot of random problems.

No Power Needed ▪ The versatile router plane excels at cleaning up dados, grooves and tenons.

Router Plane

I never knew I needed a router plane until I had one in the shop. When it comes to cleaning up tenons and grooves, the router plane is my weapon of choice. Because it has a controllable depth setting, it is user-friendly and probably the easiest tool for power-tool users to wrap their brains around. The router plane is the perfect transitional tool, bridging the gap between hand tools and power tools and setting the woodworker up for early hand-tool success.

The Cutting Edge ■ With the right medium and good technique, your tools will always be razor sharp.

My Sharpening Kit ■ I sharpen with Shapton ceramic stones, a low-grit DMT diamond plate, and a Veritas MKII honing guide.

Ceramic Stones ■ Shapton's ceramic stones last a long time and unlike traditional waterstones, require no pre-soaking.

the process is long and arduous, you aren't likely to do it as often as your tools require. But if it only takes a few seconds to refresh an edge, you won't hesitate to pause and sharpen up whenever you notice a tool is becoming dull. The longer you let your tools go between sharpenings, the more work you'll need to do to repair the damage. So the key to doing less sharpening is to sharpen more often. Really.

With so many sharpening systems in existence, from something as simple as high-grit sandpaper on plate glass to expensive motorized wet grinding stones, choosing a system to start with can be daunting. Keep in mind the fact that all these systems do the exact same thing: They abrade a metal edge with progressively finer grits. The medium that does the work and the form factor used to do it are really the only two variables. I have experimented with just about every sharpening product on the market and I have settled upon a system that works quite well and will stand the test of time. I should also point out that it isn't cheap, but keep in mind there are many brands and alternatives that will allow you to have a similar setup on a budget, though there will be a few compromises here and there. This is the system I use for most of my chisels and planes. If a tool mentioned in this section requires special treatment, we'll discuss it accordingly.

Waterstones

My system is based on waterstones. There are lots of great stones on the market including some that sport two different grits, one on each side. These can be a great choice for budget-conscious woodworkers looking to get the most bang for the buck. The stones I use are Shapton ceramic waterstones. They differ from traditional waterstones in that they are more wear-resistant and they don't require a long soak time before use. Waterstones come in a wide range of grits but I only use three: #1,000 grit, #5,000 grit, and #8,000 grit. You can think of these as medium, fine and very fine. Because all waterstones eventually wear with use and become slightly dished out, you also need a method for keeping them nice and flat. Additionally, when tools are in really bad shape you also need a very coarse grit to reshape the edge. For both tasks, I rely on a DMT DuoSharp plate. At more than $100, this diamond plate is not cheap, but it lasts a very long time. It is coarse enough to handle any minor edge reshaping, and large enough to use as a flattening plate for my three waterstones. Simply rubbing the stones against the plate after every use keeps them flat and clean.

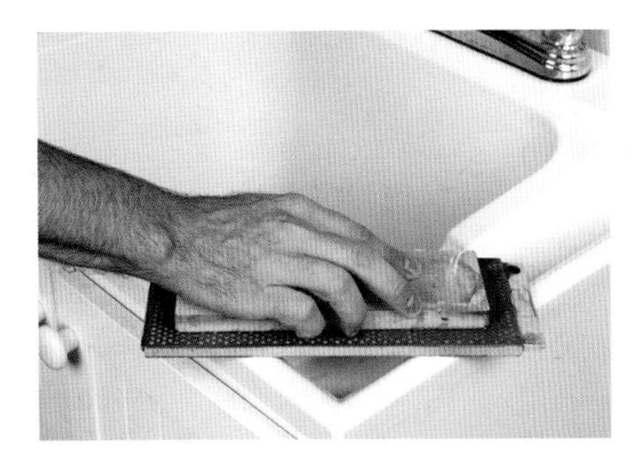

Flattening the Stones ▪ Waterstones become dished after use. At the end of every sharpening session, I rub the stones on an extra-coarse DMT diamond plate. When the stone is clean, it's flat.

Honing Guide

The final item I recommend is a honing guide. A honing guide is a small jig or carriage that securely holds the blade at a fixed angle. Some woodworkers are adamantly against them but I find honing guides to be very useful, especially for the new woodworker. Holding a chisel or plane blade at a precise angle for an extended period of time can be a challenge. If you don't mind putting in the time to practice your honing technique as well as maintain it through the years, then freehand sharpening might be a fun and useful skill to learn. Personally, I prefer to spend as much time as I can becoming a better woodworker, not a better freehand sharpener. Sharpening is a means to an end and a honing guide is nothing more than a shortcut to that end. To me it makes good sense. But don't let that deter you from trying to learn freehand sharpening if the mood should strike you. The honing guide I use is the Veritas MK II; it is the Cadillac of honing guides with a lot of bells and whistles that I rely upon, but many folks do get buy with a basic $12 honing guide.

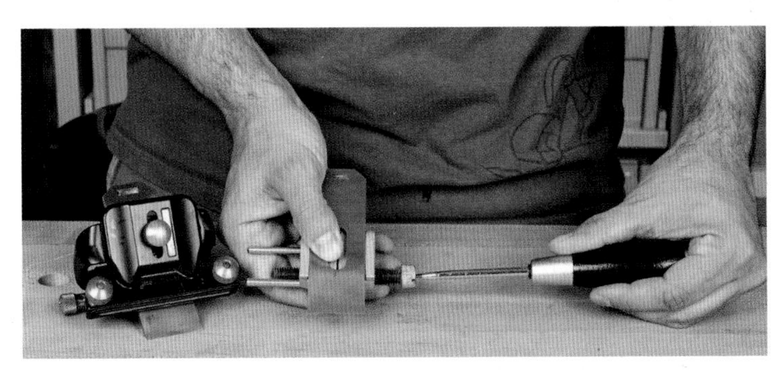

Guides for all Budgets ▪ The Veritas MKII on the left is the Cadillac of honing guides, but the $12 honing guide on the right will certainly get the job done.

While each tool blade may have specific nuances that varies the sharpening process, the principle is always the same: The back and bevel must both be flat. The line where these two surfaces meet is the cutting edge and the finer we hone and polish these surfaces, the sharper the edge. So everything we do will be toward creating and maintaining that sharp working edge.

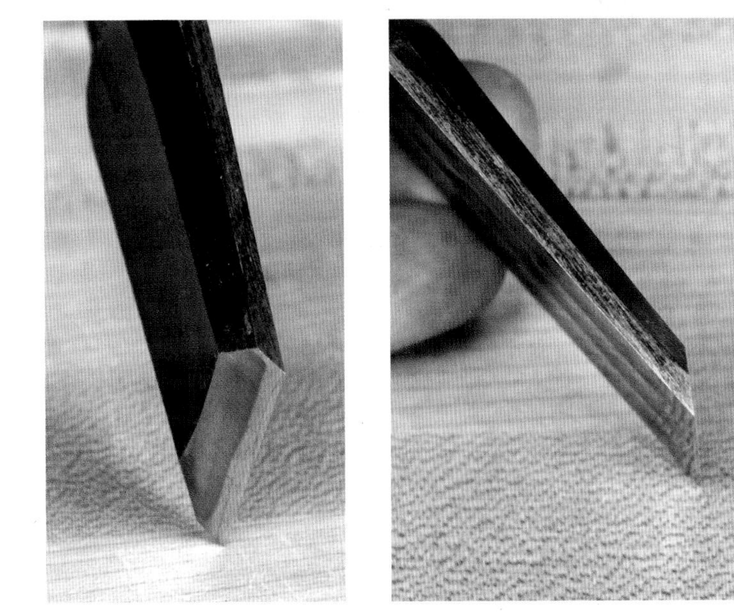

What is Sharp? ▪ A sharp edge is the line where two flat planes meet. In this case it's the plane of the ground and honed bevel, and the plane of the chisel back.

Flattening the Back

The sharpening process begins by flattening the back of the tool. Flattening the entire back of a blade, whether a chisel or a plane iron, can be difficult and time consuming. Thankfully, in most cases, it is only necessary to flatten the leading inch or two because this is the surface that is used for reference when woodworking, and you only need to do it once. The lapping process is quite simple and it is always done freehand. Hold the back of the blade flat on the surface of the stone with the handle in one hand and the other hand applying downward pressure onto the blade. Slide the blade forward and back, trying to cover the stone evenly and focusing on the outer inch or two of steel. If the blade is brand new and pre-lapped at the factory, you can usually start with a fine grit such as the #5,000 waterstone, and work your way up to a fine polish. If the blade is well-used but new to you, test it on a very fine stone. After 10 to 20 strokes, inspect the back and see how close the newly polished area comes to the edge. If the surface is uniformly polished up to the edge, the back is flat and you can proceed with the polishing process to #8,000 grit. If the back looks uneven, you may need to drop down to the lowest grit to start the flattening process from square one. Take your time with each grit and work the surface until the scratch pattern is uniform and covers the entire area up to the edge. If this sounds like a pain, that's because it is. Fortunately, once the back of a blade has been flattened, it almost never needs to be re-flattened again. We only need to maintain the flattened surface with a little polishing at the highest grit. So this is a good one-time investment of effort that you should make with each and every one of your hand-tool blades.

Flatten the Back ▪ A flat back is the first step in producing a razor-sharp cutting edge.

A Mirror Polish ▪ For a mirror shine, hone the backs of chisels and plane irons to #8,000 grit.

Honing the Bevel

With the back flat, turn your attention to the bevel. Bevel angles are tool and function-specific, so be sure to research the proper angles for your particular needs. Mount the blade into the honing guide and set it for the appropriate angle. Just as with the back of the blade, work

Hone the Bevel ▪ The honing guide helps hone the bevel to #8,000 grit.

the bevel thoroughly on each grit until the scratch pattern is uniform and reaches the very edge of the bevel. If you stroke the back of the blade, you should be able to feel a burr developing. The burr is a good sign and tells you that the bevel is truly being sharpened right up to its very edge. The metal is actually folding over and accumulating on the blade's backside. Continue working your way up to #8,000 grit. To remove the burr, flip the blade over and polish the back on the #8,000 grit stone. Hone the bevel one more time and once again remove the resulting burr, however slight. At this point the blade should be sharp and perfectly capable of shaving hair off your arm, should you have any to shave. Once you have established the bevel, you can drop the tool into the honing guide for a quick refresher any time it's needed. If you hone frequently, you'll never need to use anything but your finest-grit stone to refresh the bevel and remove the burr on the back. If you wait longer between sharpening sessions, you may need to start working the bevel at #5,000 or even #1,000 grit and work your way back up. If the edge gets damaged, you'll need to start down at the coarsest grit or perhaps consider using a bench grinder to re-shape the bevel.

A Shiny Bevel ▪ Hone the bevel just like the back, until you can see yourself in the reflection.

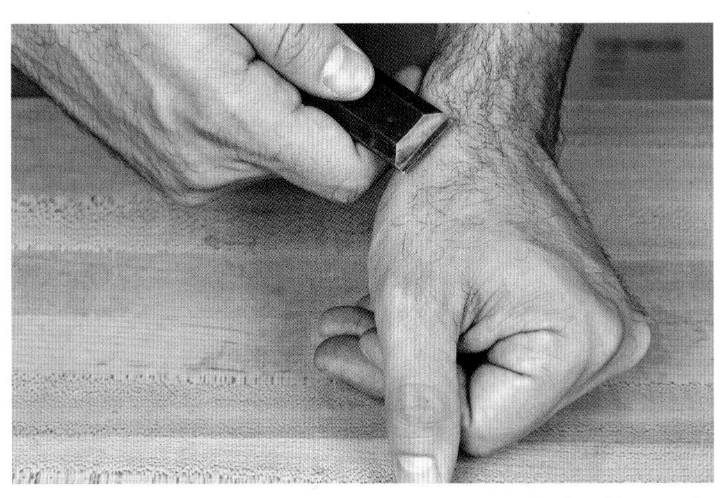

Sharp Enough to Shave ▪ While I honestly don't recommend doing this, a good, sharp blade should easily shave the hair from your arm (if you have arm hair).

Micro-Bevel

A common time-saving technique is to create a micro-bevel at the very tip of the blade. After honing the bevel, tip the blade up by a degree or two and continue honing. If you're using a guide like the Veritas MK II, the micro-bevel setting is easily engaged with the turn of a knob. It shouldn't take long to establish a new bevel at the very tip of the primary bevel. It shouldn't be more than about 1⁄16" wide. Finish it off by removing the burr from the back. So what's the purpose of this micro-bevel? Faster honing. The more metal you need to abrade when sharpening, the longer it takes. If all you have to do is sharpen a tiny micro-bevel, you can hone the cutting edge to perfection in a matter of seconds, and if you are using a good honing guide, the process is both fast and repeatable. As an example, when I notice a chisel becoming

The Micro Bevel ▪ To speed up sharpening, create a slight micro-bevel at the tip of the cutting edge. This smaller bevel reduces the amount of metal that needs to be honed.

dull, I set the guide for the proper angle, engage its micro-bevel setting, and quickly give the micro-bevel 10 passes over the #8,000-grit stone. I remove the blade from the guide, hone the back for 10 passes, and just like that I'm back to work. Micro-bevels might not be appropriate for every tool in every situation, so be sure to research the topic before you put one on all your blades.

My sharpening system is not inexpensive, but it is fast and accurate. By using micro-bevels, Shapton stones and a quality honing guide,

Other Sharpening Options

There are a lot of options when it comes to sharpening. Every year there's a new gadget or gizmo that promises to ease all your sharpening woes. Each has its own advantages and disadvantages, though most are nothing more than a variation on an existing method. It boils down to personal preference, so let's look at some other options in sharpening equipment and techniques.

Freehand

You can't argue with the simplicity of the freehand method. Just grab the blade, drop it onto the sharpening medium, and go to town. The freehand sharpening method gives a great deal of freedom to those willing to learn it. Imagine you're in the middle of a project and you realize your chisel is dull. Using the freehand method, you can re-hone and get back to work in just seconds. Also keep in mind that if you become too dependent on honing guides, you might find yourself in a pickle. Many chisels and tool blades don't fit into a honing guide, and you'll have no choice but to hone freehand.

Wet Grinder

Bench grinders are good for establishing a new bevel angle and repairing damaged edges, but the coarseness, speed and high heat make them inappropriate for fine-polished edges. Slow-speed wet grinders are a different beast altogether. The large, fine-grit stone rotates at a slow speed while passing through a small trough of water. The water keeps the tool edge cool while the stone does its work. A jig keeps the tool in position for consistent results. This machine also has a leather wheel for stropping.

Before adopting a wet grinder as your sharpening system, be aware that they create hollow bevels. The bevel of the blade is ground to match the round shape of the grinding wheel and the result is a slightly concave surface. If you decide to do some fine polishing on a stone after grinding,

Go Freehand ■ If you can get the hang of it, freehand sharpening is a great skill to have.

Wet Grinder ■ While expensive, a wet grinder quickly puts an effective edge on a tool. It has all the advantages of a grinding wheel with the bonus of staying cool, thanks to the trough of water.

sharpening is no longer a process I dread. I actually look forward to the brief diversion and it gives me incredible satisfaction to know that my cutting edges are always as sharp as they can be. The topic of sharpening is a deep one and entire books have been written on it. I encourage you to do some research and explore your options, but always keep in mind the singular ultimate goal of attaining a sharp edge. How we get there is much less relevant than the fact that we do indeed get there.

So Scary ▪ The "scary-sharp" method wins the "best value" award because it requires nothing more than wet-dry sandpaper and a flat surface.

A Woodworker's Best Friend? ▪ Diamond plates are pricey, but they last a very long time. I use mine only for coarse work and stone flattening, but there's no reason you couldn't use diamond plates as your sole sharpening system.

this hollow can help stabilize the bevel during free-hand sharpening. I find it easier to balance a bevel that has two small contact points, versus trying to keep a single flat surface in constant contact with a wet sharpening stone; your mileage may vary. On the other hand, if you're a Japanese chisel user like me, hollow bevels are generally avoided because they can weaken the cutting edge due to the layered manufacturing process of Japanese steel.

Scary Sharp

The "Scary Sharp" sharpening system gets two awards: coolest name and least expensive. It relies on nothing more than wet-dry sandpaper and a flat surface. Most folks use plate glass or granite tile. The sandpaper could be attached to the flat surface with a spray adhesive, although I find water alone works quite well. The big limitation here is the lack of sandpaper grit options. Most wet/dry paper tops out at #2,000 grit. For folks who like to polish their edges to #5,000 grit or higher, this could be a deal breaker. But there's no doubt that a serviceable edge can be achieved on a #2,000 grit surface.

Over time, you may tire of buying new sandpaper because what initially seemed like a cost-effective option morphs into an ongoing expense. So keep in mind that while there is an initial savings, the long-term cost could be the deciding factor.

Diamond Plates

While I use a low-grit diamond plate for flattening waterstones and re-shaping damaged bevels, I don't use it as a primary sharpening medium. But the truth is, there's no reason I couldn't. Diamond sharpening plates work like waterstones and are available in a wide range of grits. The higher-quality products, such as those made by Norton and DMT, will cut consistently and last a long time. The plates require no flattening and are easy to maintain, requiring nothing more than a quick rinse after use.

After trying just about every system available, I always gravitate back to waterstones and a honing guide. That's what works for me. Over time, you'll find your comfort zone too, and as long as it gives you quality results, who can argue?

Bench Chisels ▪ My primary bench chisels are of the Fujihiro Japanese variety. I like the way Japanese chisels feel in my hand and they can take quite a beating.

Stopped Dados ▪ Chisels come in handy for many things, including squaring up joints made by the router, such as this stopped dado in a cabinet side.

Bench Chisels

What tool kit is complete without at least one set of good and sharp bench chisels? None, would be the answer. Every project that exits my shop was at some point or another worked with a chisel. Although chisels come in many shapes, styles, sizes and price ranges, they all have the same basic structure: a wood or plastic handle attached to a steel blade that's sharpened on its end.

Chisels are the versatile workhorses of the woodshop and their functionality and behavior can be altered simply by changing the bevel angle. If you plan to do a lot of paring strokes, peeling wood from the surface at a very low angle, you want to keep the bevel at 15 to 20 degrees. Because the steel is thin at the tip, a low bevel angle is delicate and easily damaged. So for general use and making joints, where the chisel may take a beating, most folks want the bevel at 25 to 30 degrees. And if you know you're going to beat the heck out of the chisel for chopping mortises or general rough work, you might be best served by a 35-degree bevel. It's amazing that all of these functional variations result from a small change in the tool's geometry.

While you certainly could get by with only one set of chisels, I recommend working your way up to two sets. That's because the bevel angle can make all the difference in how the tool functions. I don't know about you but I don't enjoy changing bevel angles on a whim. So it's helpful to have chisels of all sizes dedicated to particular tasks.

How I Use Bench Chisels

The chisel is as fundamentally important to the woodshop as the saw, and its list of uses is difficult to quantify. The more comfortable you are with chisels, the more uses you'll discover. Some of the most common things I do with chisels include finessing just about every type of joint, squaring up mortises made by a router bit and cleaning dried glue squeeze-out. The versatility of a sharp chisel knows no bounds, with one big exception: Please don't use it to pry things open. My wife once used one of my chisels to open a can of paint. I get the chills just thinking about it.

Bench Chisels in My Tool Chest

I currently own three sets of chisels, if you can believe it. Set One is my beater set, the chisels I use for anything non-critical. I generally don't use them for cutting joints but they are nice to have around for general DIY work and shop projects. This set should be fairly inexpensive. I use an older set of Marples as beaters and I have their bevels sharpened to 30 degrees.

Set Two is the nice set. These are the chisels I use for finessing joints and other fine-tuning tasks. These beauties need to be razor sharp and they should be of high-quality steel that retains an edge for a good, long time, although I must confess that I do baby mine. Your budget is truly the limit here but you shouldn't have to break the bank to acquire a good functional set. I really enjoy using Japanese chisels and I currently have a set of Fujihiro chisels sharpened to 25 degrees.

Set Three is a set of mortising chisels. The thick steel body of a mortising chisel allows it to really take a beating. Whether you are a hand-tool user or a hybrid woodworker, mortising chisels make it much easier to create or finesse a wide range of mortises in a variety of soft and hard woods. The set I own is from Lie-Nielsen Toolworks and we'll discuss the use of mortising chisels a little later, in the Hand Tools to Consider chapter.

If you are new to the craft, I certainly can't recommend you go out and buy multiple sets of chisels all at once. Instead, make your first set the inexpensive set. Learn how to use them. Learn how to sharpen them. Eventually, you'll become ready to invest in a really nice set and you can move this first set down to beater status. By this time, you'll have a much better idea of your personal preferences as well as what your work dictates, thereby ensuring you get the chisel set you really need. Buying the nice set of chisels too early in your woodworking career could wind up costing you extra money if you eventually discover you don't really like the style/brand you purchased.

Inexpensive but Sharp ▪ This inexpensive set of Marples chisels holds a fine edge and is capable of tuning up joints.

Mortising Chisels ▪ Thick-bodied mortising chisels are built to take a beating.

Fine Chisels for Fine Work ▪ I reserve my Japanese chisels for the finest and most visible work.

Many Ways to Strike a Chisel ▪ Different mallets in both wood and brass are all good for hammering chisels.

Hammers and Mallets

A lot of chisel work is done using hand and body pressure only. Often times, I'll line my chisel up and simply lean into the work with a tight grip on the chisel handle. For many situations, this works well and gives me enough control to get the job done accurately. But some situations need a little more persuasion. Because sharp chisels don't need a whole lot of help even in the densest hardwoods, persuasion could very well come from a thick piece of hardwood scrap. In fact, I have a wooden hammer from my grandfather that is nothing more than a chunk of wood with a dowel handle. But if you want to fancy it up a bit, there are options.

I like having both a wood mallet and a brass hammer. While there are much nicer wood mallets on the market and I can surely make my own, Grandpa's ghetto hammer gets the job done well enough. Generally I try to use a wooden mallet on wood-handled tools, but there are always exceptions. Occasionally, when using wood-handled mortise chisels, I feel the need for a little more power, and that comes from a brass hammer.

In the brass hammer department, I once had a moment of weakness. I was taking a dovetail class in Santa Rosa, CA, and I was under the influence of Woodcraft. I had a chance to buy a set of these amazingly beautiful brass hammers from Glen-Drake Toolworks and I didn't hesitate. These brass beauties are the Cadillac of hammers, with a Cadillac price. Considering I'm on a Chevy budget, my wife wasn't all that thrilled with me that day. Ten years later the hammers are still in great shape but I have come to realize that I could have the same level of performance from a journeyman's brass mallet sold for half the money by Lee Valley.

Wood on Wood ▪ Use a wood mallet on wood-handled chisels.

Brass on Wood ▪ Japanese chisels feature a metal ferrule that prevents the handle from splitting with heavy blows from a brass hammer.

Router Plane

The router plane could be the most important tool in the hybrid woodworker's tool chest. A strong argument could be made for making a router plane one of your very first tool purchases, because it is equally useful to hand-tool woodworkers as it is to hybrid woodworkers. Oddly enough, it wasn't until I had been woodworking for several years that I came across a router plane for the very first time. My initial reaction was, "That's dumb, who would want a manual router when you can spin a bit at 20,000 rpm?" What I didn't realize at the time was that the router plane has incredible potential as a fine-tuning tool. What makes the router plane so special is that it actually does act somewhat like a powered router in that it makes repeatable cuts at a fixed depth. So unlike most other planes, once a router plane reaches a particular depth, it stops cutting. This is powerful stuff and a comforting feature for woodworkers venturing over from the world of power tools.

Many new woodworkers are intimidated by the freehand nature of hand tools because they typically don't come with measuring tapes, fences or depth stops. So a tool like the router plane really lowers the barrier to entry simply because of its controlled depth feature. Success with this tool will build confidence and encourage the woodworker to venture into other hand tools that may require more in the way of manual dexterity. While it isn't a new tool by any means, the router plane is gaining popularity among modern woodworkers of all types. Once you understand what it can do and how to use it, you'll have a very powerful ally on your quest to woodworking greatness.

The router plane is simple in design featuring a wide base with a hole in it, two round handles, and an L-shaped blade that protrudes through the hole. The look of this plane earned it the traditional nickname "old hag's tooth"; my apologies to any old hags who may be offended by this terminology. The blade is held to the primary post with a collar and the depth is adjusted with a simple turn of a knob. The plane

Consistent Grooves ■ The router plane does an excellent job of cleaning up machine-cut grooves whose depth might be inconsistent.

Precision Depth Adjustment ■ Router planes feature a depth-adjustment knob that allows you to move the cutter a precise distance below the base.

Just Enough Visibility ■ A router plane has an open area where the blade protrudes, allowing you to see what you're doing.

Extend the Base ■ A router plane becomes even more versatile when attached to a wide auxiliary base made from ½" plywood.

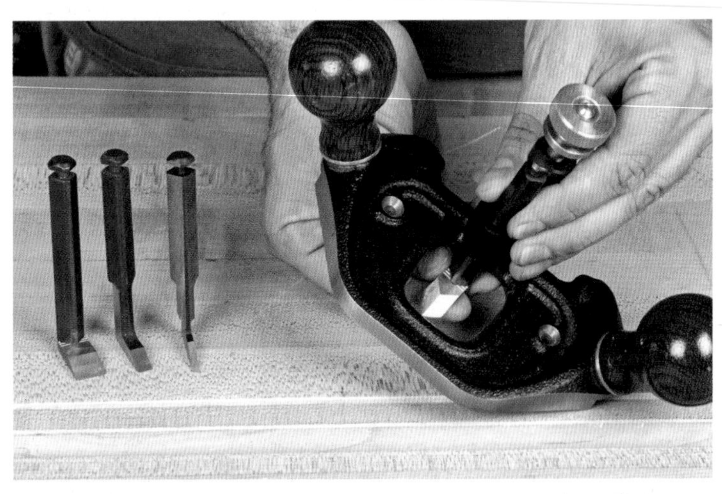

Blade Options ■ Most router planes have several different blade profiles. Some brands, such as Veritas, also feature extra-small blades.

can use blades of various widths and shapes, depending on the task at hand.

How I Use the Router Plane

One of my favorite tasks for the router plane is trimming tenon cheeks. Unlike most other planes, the router plane features a depth stop. So as long as you trim the tenon from both sides, it will always remain perfectly centered on the workpiece.

You'll never catch me cutting a dado or rabbet by hand; the table saw is generally my tool of choice for such tasks. But due to the occasional non-flat board and the tendency for boards to lift while being cut with a dado stack, many dados and grooves are not full-depth. A pass with the router plane quickly ensures that every dado and groove is of a consistent depth. Also consider what happens when you need to make a stopped dado. While a powered router is my weapon of choice for that, you certainly could use the table saw if you stopped short of the end of the dado. Because the dado blade is circular, it leaves a small ramp at the end of the cut and a router plane is a great way to clean it up.

The router plane is also great for cleaning the bottoms of shallow excavations such as hinge mortises. It can be tricky and unsafe to balance a powered router on the edge of a door stile, but a router plane does the job with grace.

Router Planes in My Tool Chest

I own a large Veritas router plane and it has given me years of great service. Numerous blade sizes are available for this particular model in both Imperial and metric, but I find I only need ¼" and ½" for most tasks.

Sharpening a Router Plane

The router plane blade is an oddity in the sharpening world. The shape of the blade makes it impossible to fit into a honing guide and the only way to get the job done is freehand. Fortunately, it's not as hard as it looks.

Start by sharpening the bottom of the blade. Place one finger on the bevel and apply downward pressure. With your other hand, grip the blade close to the stone to help maintain a low center of gravity and begin pushing and pulling across the surface. I find that the pull stroke gives the best result so I typically pull, lift, reset then pull again. It shouldn't take long to hone the surface to a fine mirror polish.

To polish the bevel. you'll need to accommodate the shaft of the blade. Either raise the stone up a few inches using scrap, or simply position the stone near the edge of the workbench. With one finger applying downward pressure and the other hand controlling the motion, skew the blade and begin pushing and pulling across the surface. Once again, the pull stroke may provide the best result and you're aiming for a mirror polish. To remove the resulting burr, rub the back on the highest grit before calling it done.

One advantage of the Veritas router plane is that the bottom part of the blade detaches from the shaft. Using the manufacturer's jig, it's easy to hone the back and bevel. While not every blade can use this jig, it is convenient for the ones that do.

Another aspect of sharpening a router plane is parallelism. Even though a blade may be honed perfectly flat on the bottom, that doesn't mean the bottom is perfectly parallel with the base of the plane. If it isn't, the blade will always preferentially cut on one side or the other. To fix it, use two equal-thickness guide blocks to raise the plane up off the workbench. Slide your desired sharpening stone in between the guides and lower the blade until it just makes contact with the stone. Move the plane back and forth a few times and check your progress. Lower the blade as needed but be careful not to gouge the stone. With just a few passes you should notice that the blade is wearing unevenly near the tip. Continue sharpening until the newly honed surface crosses the entire width of the blade. Once you establish a new flat on the bottom of the blade, you'll have to use this method again for future sharpening to avoid accidentally honing the old surface.

Different, Yet the Same ■ The oddly shaped router plane blade is sharpened just like other blades by honing the back.

Hone the Bevel ■ Get clearance for honing the bevel by propping up the sharpening stone on a scrap piece of wood.

New Advancements ■ The Veritas router plane has a detachable cutterhead that fits a small straight extension, making it a breeze to hone.

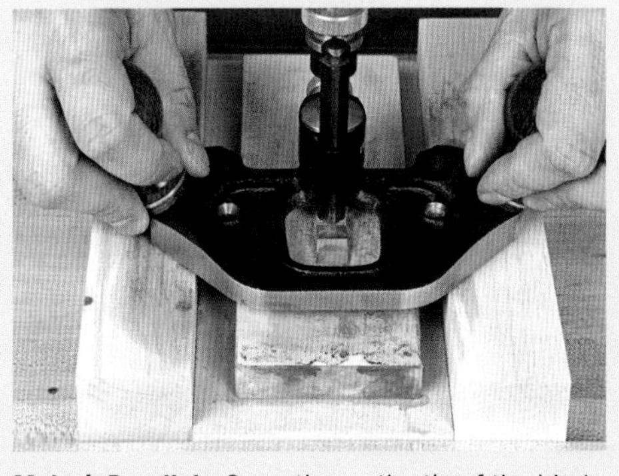

Make it Parallel ■ Sometimes, the tip of the blade is not parallel with the sole of the router plane. Remedy this by honing the back with support on each side.

Shoulder Surgery ▪ A shoulder plane, with its full-width blade and narrow body, is perfect for cleaning and squaring up tenon shoulders.

Medium or Small? ▪ Go with the biggest shoulder plane you can get. Even on small shoulders, the large model is easier to use and more comfortable in the hand.

Shoulder Plane

The shoulder plane is well suited for several tasks, but the primary one can be inferred from its name: planing shoulders. Quite often when making tenons, slight variations in setup and technique will result in uneven shoulders. This small stretch of end grain can be a real pain to trim cleanly and accurately. If you're using a power tool for the job, it's almost impossible to make the micro-adjustments necessary to even everything out. Many times your attempts to fix the problem only make it worse. Fortunately, the shoulder plane excels at this type of cut and truly is the best tool for the job.

Shoulder planes are narrow with the largest bodies topping out about 1¼" wide. The mouth of the plane is open on both sides so the blade extends all the way to the outside of the plane body. So if you're cleaning up a tenon shoulder, the flat narrow body of the plane rests nicely against the cheek of the tenon while allowing you to focus pressure down on the end-grain shoulder. You can finesse all four shoulders one thin shaving at a time, bringing them into one continuous plane.

Shoulder planes come in various sizes but I recommend getting the largest one available. I find the heavier body easier to handle on most workpieces, and the wider the body, the more useful it will be for trimming wide surfaces such as tenon cheeks and broad rabbets. In fact, if you have a large shoulder plane, you might decide you don't need the next tool in this section: the rabbeting block plane.

How I Use the Shoulder Plane

I pretty much use my shoulder plane for finessing joinery only. Because it can cut right up to an edge, it's perfect for fine-tuning tenon cheeks, tenon shoulders, rabbets, half-laps and much more.

Shoulder Planes in My Tool Chest

I currently own two shoulder planes: a Veritas medium and a Veritas large. Both planes are well-made and a pleasure to use, but I find that the larger plane is more comfortable and easier to manage for most tasks. If buying your first shoulder plane, I recommend the larger size.

Rabbeting Block Plane

We reviewed many of the rabbeting block plane's attributes and benefits during the tool-buying discussion on page 18, but they bear repeating. A rabbeting block plane is a comfortable hand-sized plane that looks almost exactly like a regular block plane but has one key feature that sets it apart: a full-width blade. The mouth of the plane extends fully to both sides of the body, so the blade also is able to extend right to the edges of the plane body. This makes it the perfect tool for any trimming task that requires you to work right up to and directly against a perpendicular edge. The most obvious example would be trimming a tenon cheek. The plane glides along beautifully, taking a wide shaving right up to the tenon's shoulder.

Some models include another handy feature called a nicker, a small retractable cutter embedded in the plane body in front of the blade on both sides of the plane. Because most cuts performed with this plane will be cross-grain, the nicker severs the fibers at the very edge, helping to prevent tear-out and making for a very clean cut.

How I Use the Rabbeting Block Plane

It can't be denied that the rabbeting block plane overlaps with several other tools including the shoulder plane, the router plane and the standard block plane. I use it to trim tenon cheeks as well as rabbets, half-laps and other joints. If I hadn't already owned a block plane when I purchased the rabbeting block plane, I might not ever have acquired the standard block plane at all. Just about anything I can do with my block plane can be done with this tool. One difference is, many folks like to dub off the corners of their block-plane blades, to prevent them from digging into the work when taking full-width smoothing passes. You can't do that with a rabbeting block plane because sharp corners facilitate its primary job.

The multi-tasking nature of the rabbeting block plane is the primary reason for including it amongst the must-have tools, while the block plane goes to the second-tier tools to consider. Depending on your existing tool kit, your goals and your methods of work, the rabbeting block plane could be an excellent addition.

Rabetting Block Planes in My Tool Chest

I have a Lie-Nielsen rabbeting block plane. It's a high quality tool that's built like a tank and does everything I ask it to do. The nickers are a nice touch that helps ensure clean cross-grain cuts.

A Shop Favorite ▪ This rabbeting block plane from Lie-Nielsen Toolworks is one of my most-used handplanes.

Tenon Cheeks in One Pass ▪ Because the rabbeting block plane is as wide as a regular block plane, it can finesse a full tenon cheek in one pass.

The Nicker Prevents Tear-Out ▪ The circular nicker on either side of the plane slices the grain ahead of the blade to prevent tear-out.

East Meets West ▪ The Western saw at the left creates a thin kerf cut, but not quite as thin as the one made by the dozuki at right.

The Workhorse Dozuki ▪ I use the Western saws for joinery and the dozuki for everything else such as cutting ebony plugs.

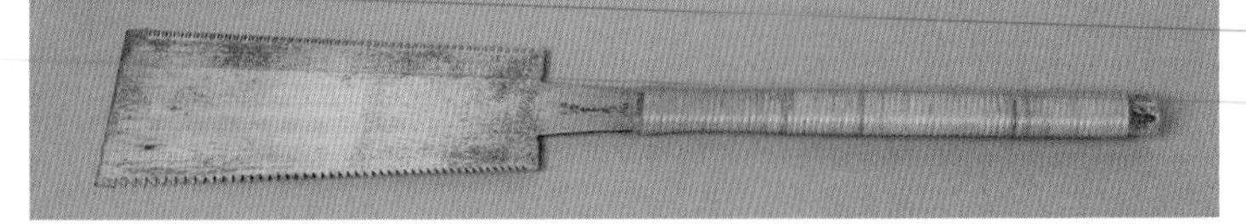

The Ryoba ▪ The Ryoba does double duty because it features a different tooth configuration on each edge. Without a reinforced back, there is no limit on its depth of cut.

Dovetail, Dozuki and Ryoba Saws

The table saw and band saw render large handsaws fairly useless in a hybrid shop, but small saws for more detailed work can be valuable. Sometimes I like to cut joints such as dovetails by hand, so I consider a small rigid saw with a thin blade and minimal set to be ideal. "Set" refers to how much the blade's teeth protrude from the flat plate of the saw body. A saw with a small amount of set can cut very thin kerfs, which is exactly what we want when cutting dovetails and other small joints. I generally recommend either a Western-style dovetail saw or a Japanese dozuki with rip teeth for this work. Both saws feature a reinforced back that stiffens the blade, but also limits the depth of cut. That's why I usually reserve this saw for smaller joints and basic trimming tasks. For many furniture parts, the limited depth is enough to get the job done.

There are a number of differences between Western saws and Japanese saws but a few of the most notable are directionality and kerf width. When using a Japanese dozuki for the first time, most Western woodworkers are surprised to find that the saw cuts on the pull stroke. If you have used Western saws routinely in the past, this may be a large obstacle to overcome. Because the saw cuts on the pull stroke and (thanks to special tooth configurations), the sawplate of a dozuki can be very thin. That brings us to the second difference: thin kerfs. While this is an asset in many situations, you need to think about the downstream effects, especially if you plan to use a dozuki to cut dovetails. Standard coping saw and fretsaw blades just won't fit into the narrow kerfs, making it difficult to remove the stock between the pins and tails. One option is to chop the waste away completely with a chisel. A second option is to invest in thinner fretsaw blades, though the thinner the blade, the longer it takes to cut. But if you prefer to use a dozuki to cut dovetails, you do have some reasonable options.

Another useful multi-tasking saw is the Japanese ryoba, which also cuts on the pull stroke. Ryobas are easy to identify because they have cutting teeth on both edges of the blade with no reinforcing backbone. What makes them so versatile is that the teeth on one edge are coarse for rip cuts while the teeth on the other edge are fine for crosscuts. Lots of folks use ryobas for sawing dovetails and other joints, too and because there is no backbone, there's no limit on cutting depth. But like any jack-of-all-trades, you might find this saw isn't ideal for every situation. Fortunately, the ryoba comes in numerous sizes and the smaller they get, the finer the teeth. So if you're looking for a finer cut, simply look at the smaller size range. For a more aggressive cut, go big.

How I use Small Saws

My small saws are used not only for cutting joints such as dovetails, but also for just about any general cutting or trimming task that comes across the workbench. Some examples include cutting dowels to length, cutting small ebony plugs for Greene & Greene work and cutting parts or joints that might be awkward or unsafe to saw with a power tool.

Small Saws in My Tool Chest

I own both a Lie -Nielsen Western-style dovetail saw and a very generic dozuki. I tend to reserve my Western saw for fine joinery tasks such as dovetails and small tenons. My dozuki is treated more like a general-purpose saw and I use it for anything and everything. The blade on the dozuki is also replaceable and interchangeable and can be converted to a flush-trim saw. I don't own a ryoba because it would be redundant when combined with the tools I already have, but if you haven't yet purchased a handsaw, a ryoba is certainly worth consideration.

Flush-Trim Saw

Early in my woodworking career, I often used screws to reinforce plywood and solid-wood case joints. The screws were countersunk in ⅜"-diameter holes and I simply glued in small pieces of dowel stock to cover their heads. After the glue dried, I was left with small pieces of dowel stock protruding from the surface. The only way I knew to make them flush was to grind them down with a sander. Unfortunately, because the dowel is essentially end grain, it would sand at a much slower rate than the surrounding material. I would end up inadvertently burning through the veneer layer around the dowel, or simply creating random hills and valleys if the surface was made of solid wood. I knew there had to be a better way and after posting a few questions in a woodworking forum, I had the answer: a flush-trim saw. Since then, the flush-trim saw has become one of my most-used hand tools and is near the top of my list of the hybrid woodworker's secret weapons.

A flush-trim saw has two important features that helps it do its magic. First, the blade is very thin and flexible. Simply lay the saw blade on a flat surface and lift the handle while keeping downward pressure on the blade. This allows you to move the blade back and forth while keeping constant contact with the reference surface. With any other saw blade, the teeth would scratch the surface and create a big problem. But the second key feature of a flush-trim saw is that the teeth have zero set, meaning the teeth don't protrude out from the saw plate. When moving the saw

Flush-Trim Saws ▪ Flush-trim saws feature a thin, flexible blade and the teeth are ground flat with no set.

Trimming Plugs ▪ I use the flush-trim saw for a number of shop tasks, but most often for trimming plugs flush to a surface.

Clean Results ■ After flush-trimming, the surface needs only a light sanding before it's ready for finish.

Unexpected Use ■ The flush-trim saw can be used in non-obvious ways, such as flushing up this proud shoulder on a large tenon.

back and forth, the only thing being cut is the wood directly in front of the teeth. Any material directly beneath the teeth remains unharmed.

So in the plug-trimming example, you can see how incredibly useful this tool would be. Simply slice the plugs flush with the surface and lightly sand the area smooth, using the random-orbit sander or a wooden block wrapped in sandpaper. Most times the plugs are almost perfectly flush after trimming but occasionally, if you accidentally lift up on the saw, you might end up with some extra material sitting proud of the surface. This is easily remedied by making another pass with more downward pressure on the blade. What a massive time-saver!

How I Use Flush-Trim Saws

The list of things I like to do with flush-trim saws might seem modest at first glance, but there are hundreds of situations where trimming one surface flush to another will be useful. As discussed above, flush-trim saws are perfect for trimming plugs flush to the surface. Just a few strokes saves a significant amount of time over sanding and yields better results.

Here's another use that might not be immediately obvious. When making tenons on the table saw, the short shoulders sometimes wind up a little proud of the long shoulders. If the shoulders are wide enough, as they might be on large tenons made in 8/4 stock, you can remedy this using the flush-trim saw. Resting the saw against the flat of the long shoulders, carefully push and pull the saw while tilting it into the offending wood. Slowly but surely, the extra material will be removed. Anything you can't quite reach with the saw can be cleaned up with a few strokes of a sharp chisel.

Another unexpected use for a flush-trim saw is in joinery. Because the saw cuts so well when pressed against an edge, you could use it with a guide to start a critical cut. Let's say you want to cut a tenon shoulder by hand: Simply clamp a block right on the shoulder line and use the flush-trim saw to cut the shoulder. Some may consider that a training-wheels approach but I call it smart. Although I don't recommend cutting tenons by hand (unless you really want to), understanding that this option is available might inspire a similar solution in a completely different situation.

Flush-Trim Saws in My Tool Chest

I own two flush-trim saws and I honestly don't know the brands. The older of the two features a wood handle and teeth on only one edge of the blade. A newer model has a black plastic handle and cutting teeth on both edges of the blade. Both models get the job done and I don't feel this is a tool we need to be too picky about.

Spokeshaves

Early in my woodworking career, I became infatuated with curves. I found myself trying to incorporate curves into everything, including places where they didn't belong. Sometimes you need to cross the line in order to find out where it is; live and learn I suppose. During that phase I learned the value of the spokeshave for shaping and finessing curves. I use two power tools to cut curves: the band saw and the router. The band saw makes quick work of the task but leaves a rough washboard surface. A flush-trim bit and a router with a template will certainly yield cleaner results than a band saw, but the process takes longer and the surface still isn't perfect. Furthermore, some shapes simply don't lend themselves well to template-routing. Enter the spokeshave. Although spokeshaves were originally designed for shaping chair spindles, I find them useful for shaping and smoothing just about any curved workpiece. Thanks to its narrow body, the tool traverses tight curves with ease.

Spokeshaves feature a narrow body, a short iron, and two long, straight handles. The tool can be pulled or pushed through the work, which is very helpful for dealing with grain changes. On most curved surfaces, the grain will change at least one time. You'll be pushing the tool with the grain when working downhill and then against the grain when moving the tool uphill. Whenever the grain changes, simply flip the tool around and work in the opposite direction so you continue cutting with the grain. This is great for lazy people like me, who would rather not have to reposition the workpiece.

Spokeshaves come in three body styles: flat, round and concave. The flat version is good for general use and outside curves. The round version is well suited for inside curves. The concave version is made specifically for spindles and other near-round workpieces.

How I Use Spokeshaves

To put it simply, spokeshaves help me finesse inside and outside curves. The surface left by most cutting tools is not only rough, but it can be difficult to approach your line without cutting too far. The spokeshave allows me to work a curve a small shaving at a time until the curve is exactly where I want it. And although the sole of the tool is small, it can help level out small divots and hills resulting in a smooth continuous curve.

Spokeshaves in My Tool Chest

I own a Veritas round and a Veritas flat spokeshave. I find these two tools cover all of my curve-planing needs and in my work, I generally have little use for a concave spokeshave.

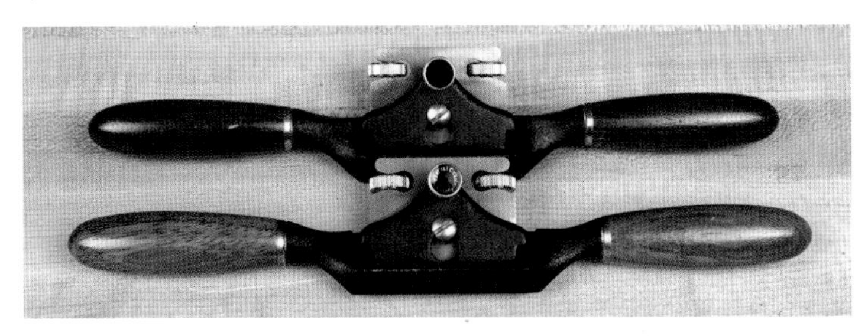

Beautiful and Useful ▪ Spokeshaves are a staple of the curve-maker's tool kit. Both inside and outside curves can be finessed to final dimension with a good set of spokeshaves.

A Subtle but Important Difference ▪ There are a few different types of spokeshaves on the market but the two I find most useful are the flat bottomed and the round.

Just a Short Plane ▪ Like most other planes, a well-tuned spokeshave takes nice, light shavings.

Grain Direction

Every board has grain direction and it's a natural feature of wood that woodworkers need to become intimately familiar with. A good analogy for grain direction is the fur on my black Lab's back. If I pet him from head to tail, my hand passes smoothly along his fur with no friction. But if I pet him from tail to head, my hand will catch in the hair making for a rougher feeling. While my dog seems to enjoy the attention either way, one direction is clearly against the grain. It's the same for wood. Run a plane with the grain and you will experience a smooth cut, free of tear-out. Run it against the grain, and you'll deal with excessive plane chatter and tear-out.

The Edge Reveals the Face ▪ To judge the grain direction on the face, look at the edge. The plane in the picture on the left is going with the grain while the plane on the right is going against the grain.

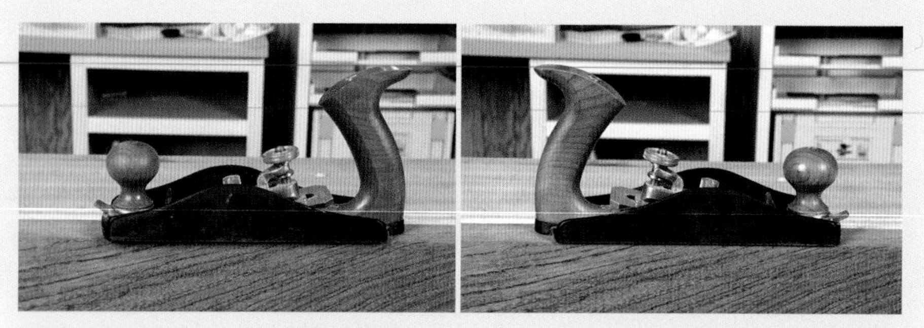

The Face Reveals the Edge ▪ To judge the grain direction on the edge, look at the face. The plane in the picture on the left is going against the grain while the plane on the right is going with the grain.

As woodworkers, it's important for us to learn how to read the grain. Fortunately, this is easy to do on most boards. If you want to know the grain direction on the face of a board, simply look at the side grain. If you want to know the grain direction of the edge of a board, take a look at the face. This methodology is important for not just handplanes, but also power tools, as you'll see when we discuss milling (page 93).

Because grain lines aren't always as evident as we'd like, some boards show little to no visual indicators of grain direction. In these situations, I employ the pantyhose trick. For the gentlemen in the crowd, there is no shame in buying pantyhose for your shop. My wife doesn't wear pantyhose so I had to buy my own and I lived to tell the tale. Simply bunch the hose up into a ball and run it across the surface of the board in both directions. You should notice that in one direction the material slides freely (with the grain), but in the other direction it catches on seemingly invisible fibers (against the grain). It's a pretty cool trick for when all other methods fail.

Pantyhose Trick ▪ It may sound crazy but pantyhose can help you determine grain direction in difficult-to-read woods.

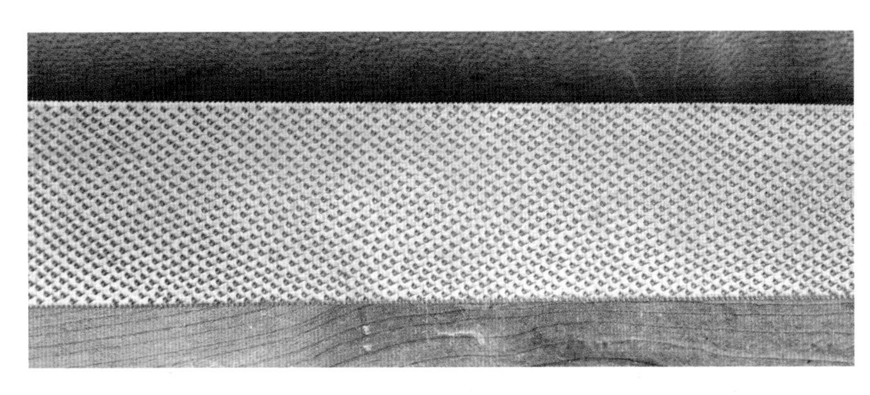

Not All Rasps are Equal ■ A high-quality hand-stitched rasp vastly outperforms a machine-stitched version.

Rasps

Previously, I mentioned my unhealthy love of curves and a tool that makes curved work not only possible but also fun is a rasp. A rasp looks like a metal file, but if you look closely at the cutting surface you'll notice a dramatic difference. Instead of regular repeating grooves, there are lots of tiny metal teeth that are the result of a process known as stitching. Essentially, a metal punch dents the surface of the metal, causing it to bulge up into a sharp tooth. Rasps come in various shapes, sizes and degrees of coarseness. Not all rasps are created equal so if you plan on doing a great deal of curved work, invest in a variety of high-quality hand-stitched rasps.

Rasps are always used freehand, so it will take a little practice to build your confidence in using them. Fortunately, unlike handplanes, you usually can't do much damage with one errant stroke. Depending on how you hold the rasp, where you apply pressure, which part of the rasp is contacting the wood and whether you go with or against the grain, you can significantly impact the resulting cut. If you're completely new to the concept of rasps, it's a good idea to put a chunk of scrap wood in your vise and have at it. After a few minutes of fun, you'll have a much better sense of what this tool can do for you. If you're intimidated by the fact that rasps don't come with fences or guides to control the depth and direction of the cut, don't be. I almost never use a rasp without having some sort of guideline in place for reference. Not being a particularly artistic person, I can't always go by eye, but I find, that drawing reference lines turns an artistic process into a systematic one that yields artistic results.

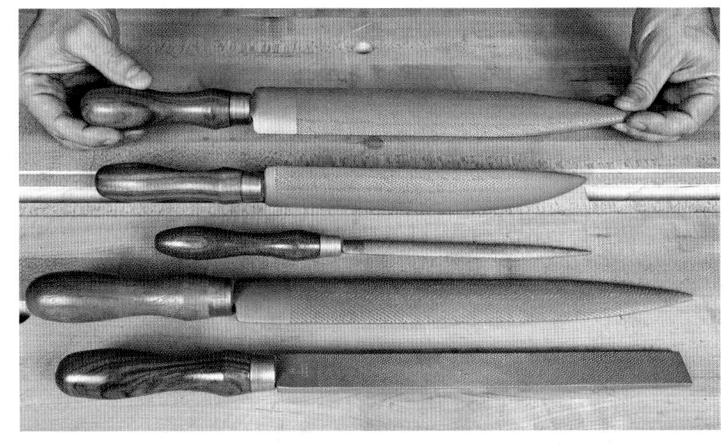

A Nice Variety ■ If you do a lot of shaping, you will want rasps in a variety of sizes, shapes and degrees of coarseness.

Not Just for Looks ■ A rasp makes short work of rounding square tenons so they fit in round-ended mortises.

How I Use Rasps

One of the best features of a rasp is that it can be used just about anywhere without having to worry much about grain direction. If you're looking to get into sculpted furniture, such as Maloof-inspired chairs, the rasp will quickly become one of your best friends as you abandon the rigid world of straight lines and embrace the freedom of sculpture. I also like to use rasps to create roundovers and other edge treatments on workpieces that can't be worked with a router.

When I make mortise-and-tenon joints with power tools, I wind up with square tenons and round-ended mortises. With just a few strokes of the rasp, I can round over the corners of the tenons so that they fit the round mortise perfectly. This is much easier and faster than squaring the mortise with a chisel.

Rasps in My Tool Chest

I own numerous rasps of various shapes, sizes and degrees of coarseness, because I do a lot of free-form shaping. Not everyone will need such a wide variety. For the average shop, I recommend three: a large coarse rasp, a large fine rasp, and a small fine rasp. The two larger rasps will handle your heavy and light stock removal on most workpieces and the smaller rasp comes in handy for tight spaces and other odd jobs. I recently upgraded all of my rasps to the high-quality (and high-priced) hand-stitched Auriou rasps. The initial purchase did hurt, but the pain subsides with each use.

Sculpted Chairs ■ If you ever plan on making sculpted furniture, such as a Maloof-inspired rocker, you'll need to become comfortable with the rasp.

Reference Lines And Rasps

Sculpting curves with a rasp may seem like a purely artistic endeavor. There's no bearing, fence or stop preventing you from removing too much wood so it really is up to your manual dexterity. But that doesn't mean it requires a high level of innate ability. Most of us are capable of removing wood to a line using various other tools, so using a hand-held rasp should be a piece of cake. And although there is some skill involved, it truly is a systematic process that yields artistic results. If a non-artistic person like me can do it, so can you.

Let's create a large roundover in a piece of 8/4 cherry. Normally I wouldn't bother using a rasp to make a roundover, but the exercise will show you that by using simple guidelines you can use a free-form hand tool such as a rasp to end up with a surface that is clean and consistent. You can apply this method to all types of sculpted furniture parts.

The first step is to draw the guidelines. On each end of the board, use a fancy template (a roll of electrical tape) to draw the roundover profile. Using an adjustable square, extend the top and bottom parts of the roundover along the face and edge of the board. These lines represent that extremities of the roundover. Because we'll create the roundover by first cutting a series of chamfers, draw a second set of lines to represent the extremities of a 45-degree chamfer. Note that this chamfer should be as close to the roundover as possible without going through it.

Begin removing the bulk of the waste in the form of a large chamfer, being careful not to go beyond the inner lines. Keep the rasp locked at 45 degrees and work from one end of the board to the other, keeping an eye on the end markings to gauge progress. Next, create two smaller chamfers connecting the first chamfer to the outer lines. The lower chamfer will require the rasp to be nearly vertical. The upper chamfer will require the rasp to be nearly horizontal. With three chamfers on the corner, the profile approaches round, even though it's just a series of flat facets that approximate the desired shape. At this stage the roundover is close enough that the rest of the work is done by eye. Work across the edge knocking down the high points and matching the profile drawn on the end of the board. Before you know it, you have a near-perfect round-over made completely by hand.

I typically use routers to make roundovers, but the example demonstrates the foundation on which you can construct more complex shapes and curves. Once you have a few reference lines in place, all you need is a systematic approach to removing the material that sits between you and your curvy creation. This process is not limited to rasps, either. You can employ the same systematic methodology with spokeshaves, handplanes, ball mills and heavy-duty carving blades installed in a grinder.

Step 1 ▪ Mark the profile on each end.

Step 2 ▪ Extend guidelines down the length of the face and edge.

Step 3 ▪ Extend another set of guidelines for the tangential 45-degree mark.

Step 4 ▪ Create the first facet, a simple 45-degree chamfer.

Step 5 ▪ Use the inner set of guidelines to gauge progress and don't go past the line.

Step 6 ▪ Create the second facet by tilting the rasp forward and removing the wood back to the second pencil line.

Step 7 ▪ Create the third facet by lowering the rasp and working to the final line.

Step 8 ▪ Blend all of the facets into a smooth and continuous roundover.

Protect Your Fingers

If you use rasps in your work, you'll soon know the misery of sore fingers that feel like they've been passed over a cheese grater. For long-term comfort, take a few steps to protect those sensitive digits. First, consider an auxiliary rasp handle. The one I use is made by Lee Valley Tools and it simply attaches to the end of any rasp. With a regular handle at one end and an auxiliary knob at the other, you can remove stock in a hurry without doing any damage to your fingertips.

A second option is more versatile because it works with any rasp and it never gets in the way: finger tape. You can buy flexible finger tape in rolls and you'd be surprised at how effective it can be. Just wrap a few pieces around your fingertips and you'll not only be protected from the rasp's wrath, you'll also have a more effective grip on the tool. For long shaping sessions, you just can't beat finger tape.

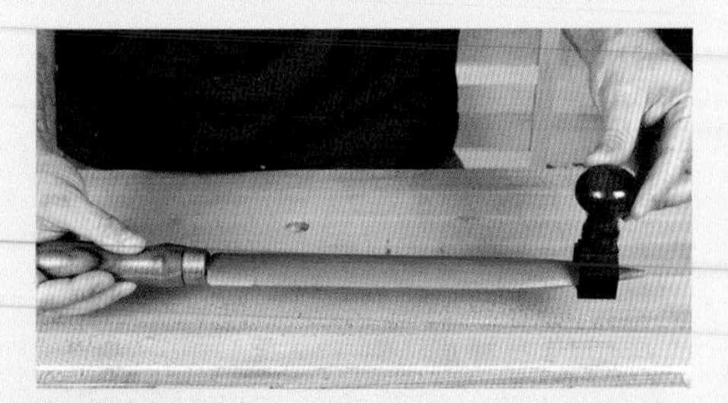

Get a Grip ▪ An auxiliary handle gives you leverage and reduces fatigue.

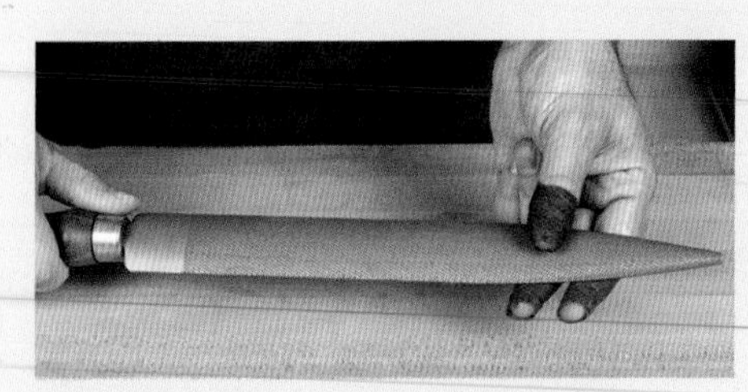

Finger Tape Prevents Pain ▪ Finger tape prevents your fingertips from becoming raw after extended rasp use.

Scrapers

I haven't met the woodworker whose eyes didn't light up the first time they saw a scraper in action. Most folks are utterly surprised to find out that they have an additional option for smoothing a surface, besides a smoothing plane and a sander. Sanding is generally unappealing due to the noise and the dust it generates. Smoothing planes require practice and they can be temperamental on dense and figured woods. But a card scraper, when properly sharpened, can deliver tear-out-free results on any wood regardless of grain or density, and it takes about five seconds to get the knack of using it.

The magic behind a scraper is its cutting edge. The steel of a scraper is drawn out using a burnisher and then curled over into a slight hook. This hook acts like a small cutting blade and allows us to peel very fine shavings off the surface of the wood. Milling marks from the planer, jointer and table saw disappear with two or three passes. Scraping is a much more pleasurable and less time-consuming experience than sanding from #80 to #220 grit with a random-orbit sander. And unlike a smoothing plane that can do significant damage when it gets dull, a scraper simply stops cutting and begins to generate dust instead of shavings.

While a scraper will certainly save you time, energy and money over sanding, don't throw away your random-orbit sander just yet. You'll have to decide for yourself if the surface left by the scraper is truly finish-ready. In most cases, I find that it isn't. Scrapers can leave minute track marks that would be visible in the final finish, so

Shavings, Not Dust ▪ A properly sharpened scraper will produce short shavings. If you're producing dust, the burr is dull or not properly prepared.

I sand a freshly scraped surface with #220 grit as a final preparation step before applying the finish.

Scrapers come in three notable forms. The first is the humble card scraper. A card scraper is nothing more than a thin piece of steel cut into a particular shape. The rectangle is the most common. It works well on flat surfaces, curved surfaces and along inside corners. Other shapes include a rectangle with one concave edge and one convex edge, and the gooseneck scraper, which is useful for curved surfaces and profiles. They also come in different thicknesses so be sure to try a few before committing to a set. I prefer a thick card scraper because I find it easier to sharpen and use, and it maintains a sharp edge longer than thinner scrapers. Card scrapers are cheap so there's no reason not to have several in the tool kit.

One issue with a card scraper is the intense heat that can build up. After five or six passes on a dense hardwood, the steel heats up enough to burn your fingers and thumbs. A quick trick to help prevent this from happening is to put a magnet the size of a business card on each side of the scraper. The magnet insulates just enough to prevent your fingers from burning and that makes the scraping experience much more enjoyable. Scraper holders are also available, and the other two scraper types keep your hands off the blade entirely.

The second form factor is the cabinet scraper, the Stanley No. 80 or similar copies. This tool features a thick blade milled with a 45-degree bevel, which makes burnishing easy, and two side handles. The cabinet scraper can take more substantial shavings than a regular card scraper and has the additional advantage of a sole surrounding the blade. This reference surface helps when leveling uneven surfaces such as panel glue-ups. An additional bonus is that your fingers won't get hot because you aren't holding the blade directly. I purchased two No. 80 cabinet scrapers on eBay about 10 years ago and I keep one tuned for rough work such as glue scraping and rough prep, and one tuned for finish preparation.

The third form factor looks more like a bench plane than a scraper and is generally called a scraper plane. This type of scraper has a few notable advantages. First, it is comfortable to hold and use for long periods of time. Card scrapers heat up and eventually your thumbs get tired. Cabinet scrapers are more comfortable but with your upper body and thumbs doing most

A Shape for Every Purpose ▪ Card scrapers come in various shapes for flat, concave and convex work.

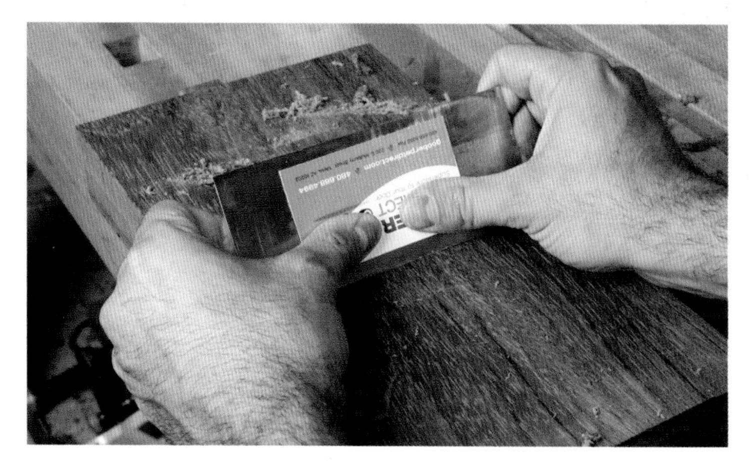

Beat the Heat ▪ Something as simple as a refrigerator magnet can insulate your fingers from a card scraper's heat.

No. 80 Cabinet Scraper ▪ The cabinet scraper features a wide base that helps promote surface flattening.

Sharpening Card Scrapers

To sharpen a scraper, we must first understand what makes a scraper different from other cutting edges. Instead of refining two faces to create a razor-sharp edge, a scraper relies on a burr or hook that is created by drawing the metal out from the edge. To do this, you'll need two additional tools, a single-cut file and a burnishing rod, as well as a simple shop-made reference block. The process begins by honing the edges of the card scraper so they are smooth and square. An old beat-up scraper usually requires filing before it can be honed on the waterstones. You can do the filing freehand with the scraper in a vise or you can get a little help from a reference block. I like to cut a groove into a block to hold the file snug and square to the surface. This way I can simply drop the block and file onto the scraper and have perfect square registration while moving the file forward and back.

Once the edges have been filed clean and square, it's time to hone them. The reference block holds the scraper perpendicular to the surface of the stone. I usually hold the block stationary on the stone surface while pushing the scraper forward and back. After the edge has been honed, lay the scraper flat and polish the outside inch of steel. This is the same exact process as sharpening regular blades. Work your way up to #8,000 grit.

With the edge honed perfectly square, we'll create the burr by drawing the metal out from the edge and then rolling it over into a hook. It's a multi-step process that does take practice. To draw the metal out using the burnisher, lay the scraper flat on the workbench so that it just overhangs the edge. Hold the burnisher near horizontal and draw it across the edge of the scraper five or six times with a decent amount of pressure. A drop of oil on the burnisher will reduce friction and heat dramatically. The next step requires holding the scraper on edge. Some people hold by hand but I am more comfortable with the scraper in a vise. Hold the burnisher about 5 degrees off horizontal and run it along the scraper's edge three or four times, once again trying to draw the corner outward. This step not only draws more metal out, it also begins to roll the metal that was drawn out in the previous step. The final step is to finish rolling the hook and give its final shape. Hold the burnisher at about 45 degrees and make another one or two passes along the scraper's edge. You should be able to feel the cutting hook with your finger and with any luck, the scraper is ready to cut so try it out on some scrap. If the first run wasn't successful, you don't necessarily have to repeat the filing and honing. Instead, begin by rolling the edge out with the scraper flat on the bench, then repeat the burnishing process.

Sharpening a cabinet scraper is a similar process only instead of working a 90-degree square edge, the blade has a 45-degree ground bevel. Like a plane blade, the back and bevel need to be honed to a nice sharp edge, freehand or using a honing guide. The burnishing process is similar to the card scraper method only it requires fewer steps. Simply lay the blade on the workbench with the bevel up and just slightly overhanging the edge. Tilt the burnisher to about 55 degrees and run it over the beveled edge four or five times. If you aren't sure about the angle, keep in mind the bevel itself is at 45 degrees so all you need to do is tilt the burnisher to your best approximation of 10 degrees more. If you want a more aggressive hook, you can work the edge a little more, applying more pressure and tilting the burnisher to 60 degrees.

Filing the Easy Way ■ A file embedded in a thick scrap block makes for easy filing at exactly 90 degrees.

Step 1 ▪ Hone the edge through the standard range of grits, using a block to hold the scraper square to the surface.

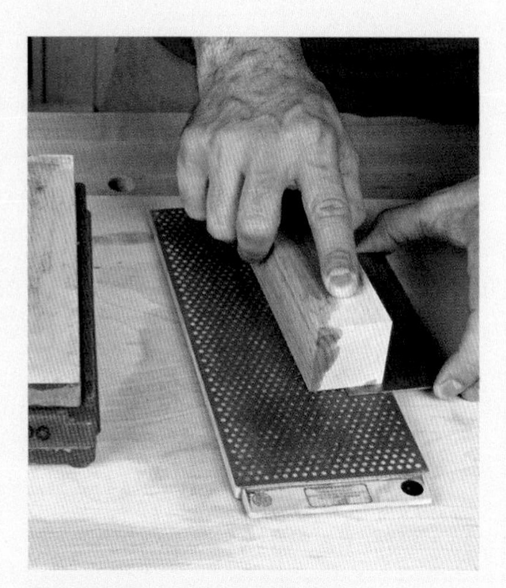

Step 2 ▪ Hone the flat side of the scraper to remove the burr. The wood block can be used to apply uniform pressure.

Step 3 ▪ After honing the edge and face, the edge should look clean and consistent.

Step 4 ▪ Burnish the edge at a 5 degree angle with the scraper flat on the edge of the bench.

Step 5 ▪ Burnish the edge 5 degrees off horizontal with the scraper in a vise.

Step 6 ▪ Burnish the edge at 45 degrees to finish the burr.

Step 7 ▪ Test the results and enjoy the sweet shavings.

Step 8 ▪ The bevel and back of the cabinet scraper blade should be honed like any other blade.

Step 9 ▪ Use the burnisher at 55 degrees with the blade flat on the workbench, bevel up.

of the work, fatigue will set in fairly quickly. A scraper plane gets around these issues by securing the scraper in a traditional plane body featuring a tote and a knob. So with your strong arm behind the plane and your legs powering the work, you can scrape more material for longer periods of time with less fatigue. This type of plane does have a couple disadvantages, though. Because it has adjustments for changing the angle of attack, it can be finicky to set up. Additionally, they are expensive. The most common models on the market today are modeled after the Stanley No. 112 and they retail starting at $169.

How I Use Scrapers

I use scrapers for leveling and smoothing uneven surfaces. As a power-tool user, I have come to terms with the fact that most workpieces will come off the tool with milling marks, chatter markd and occasional burns. Fortunately, a few passes from a scraper can relieve just about any surface flaw created by my power tools. And while all of my projects get sanded with fine-grit paper, scrapers can save me the hassle and unpleasantness of low-grit sanding.

Scrapers in My Tool Chest

No tool kit is complete without at least a few card scrapers. It's probably a good idea to also have either a cabinet scraper or a scraper plane, but not necessarily both. Both tools are great to have in the shop even though they are redundant. I own two No. 80 cabinet scrapers but I do not own a scraper plane.

Scraper Plane ■ The iron in the scraper plane has a heavy burr and sits nearly vertical. The two brass nuts adjust and lock the blade angle.

Leveling Panels ■ The scraper plane is effective for leveling glued-up panels in any wood.

Feel the Burn ■ This piece of cherry was burned by a dull table saw blade.

Scrape it Away ■ After some scraping, the surface looks smooth and clean.

Marking Gauges and Marking Knives

My good friend William Ng, owner of William Ng School of Fine Woodworking in Anaheim, Calif., once gave me a piece of advice that I'll never forget. He said, "If you want to be a good woodworker, use a pencil. If you want to be a great woodworker, use a knife." Since then, I am always looking for ways to improve the measuring and marking process by incorporating knife cuts instead of pencil marks.

The problem with a pencil line is that it has an appreciable thickness of its own. The smallest mechanical pencil produces a .5mm line, and standard pencils produce much thicker lines. So let's assume the best and go with the .5mm pencil line as an example. If you're not fluent in metric, .5mm is a strong 1⁄64". Of course in some respects, a strong 1⁄64" is very small, but in our world, it's the difference between a perfect mortise-and-tenon and a loose one, or an air-tight miter joint and one with an unsightly gap. When you mark a cutline with a .5mm pencil, you have an important decision to make: where to make the cut. Do you go in front of the line, keeping the line visible after the cut? Do you try to cut the line in half? Or do you consume the line in the cutting process? Whatever you do, you need to pick a system and stick with it, otherwise you're doomed to a strong 1⁄64" error throughout your work. Making things even more inaccurate is the fact that a pencil can never quite butt up against a straightedge. The body of the pencil itself keeps it a small but noticeable distance away from the reference surface. This is a distance you'll need to account for.

How I Use Marking Gauges and Knives

Contrast the .5mm line with a knife mark. First, the knife mark leaves an incredibly thin line. With a light touch, it has negligible thickness. In addition, the knife line essentially begins the cut that you will continue to cut, and there is no mistaking the location and the directionality of the line. Because most knives have flat backs, you can run them along a straightedge to perfectly represent your desired cut location. A knife has the additional functional advantage of severing the cross-grain wood fibers, which helps you achieve clean, crisp cuts of perfection.

Just about any sharp blade will suffice as a marking knife but it's best to avoid anything that has a bevel on both sides of the blade because running the bevel against a straightedge creates

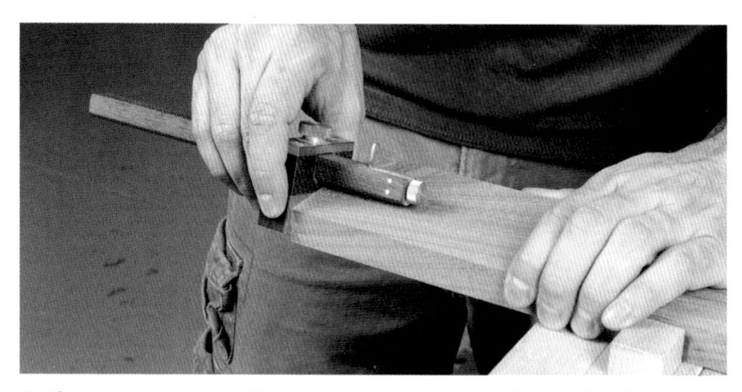

Cutting Gauge ▪ Use a cutting gauge to lay out and scribe shoulder lines, preventing tear-out during the milling process.

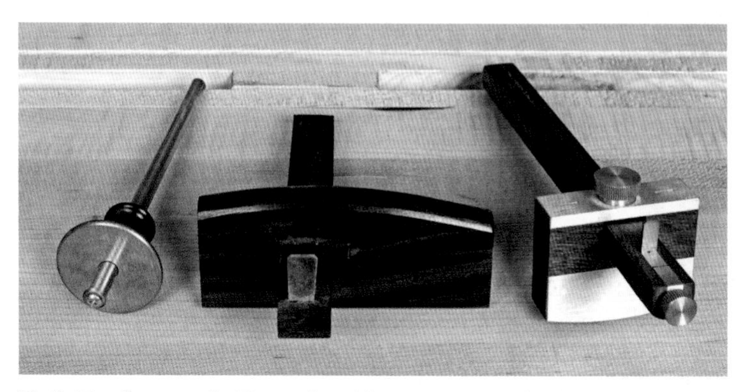

My Cutting Gauges ▪ Cutting and marking gauges come in all shapes and sizes, but they are all capable of getting the job done.

Marking Knife ▪ A marking knife not only strikes a line for reference, it also severs the grain in preparation for the next steps.

the same offset as the pencil. So for me, the best marking knife has a flat back and a comfortable handle.

A knife-style marking gauge, also called a cutting gauge, is very similar to a marking knife except its blade mounts near the end of a wooden post that is held by a moveable fence. A cutting gauge is handy for laying out tenons and other common joints. Rather than using a square and a knife, the cutting gauge allows you to scribe a cutline a fixed distance from the edge. So when you need to scribe a line multiple times or on multiple faces of a board, a cutting gauge is the way to go. Marking gauges and cutting gauges come in a variety of configurations and edges and I encourage you to experiment with them. There's a lot of room for personal preference.

I use cutting gauges primarily for marking tenon shoulders. Even though I use power tools to make the actual shoulder cuts, having a pre-scribed line ensures the shoulders will be crisp.

Jack of All Trades ▪ A jack plane does a fine job of flattening boards as well as numerous other planing tasks.

The scribe line also gives me a perfect mark for setting up the power-tool cut. Unfortunately, the marking gauge is limited by the length of its post and that's where the marking knife comes into play. Precise layout of things such as hinges and dovetails is the territory of the marking knife.

Marking Gauges and Knives in My Tool Chest

I own several cutting gauges, including a traditional knife gauge with brass fittings, a Japanese-style gauge with a wedge-secured post, and a Veritas wheel marking gauge. I purchased the various gauges so I could experience them in use. I've come to favor the traditional gauge, but I still keep the others around because I reserve the right to change my mind. My marking knife is a little thing of beauty from Czeck Edge Hand Tool, whose knives are as much fun to look at as they are to use.

Jack Plane

At this point, near the end of the Must-Have Hand Tools chapter, you're perhaps thinking, "What about bench planes?" When first learning about hand tools, some woodworkers immediately assume they need to purchase an entire set of bench planes. After all, what woodshop is complete without a beautiful display of handplanes on the wall behind the workbench? In fact, one of the most common questions I receive at my website is, "Which bench plane should I buy first?" No one ever asks the question they should be asking: "Which hand tool should I buy first?" The truth, at least for the hybrid woodworker, is that many of those big, beautiful and expensive planes are unnecessary and add little value to the hybrid tool kit. Unless you have a desire to flatten, joint and smooth boards completely by hand, save your money and let your jointer, planer and table saw do the work they are intended to do.

The one bench plane I have included in this section is the jack plane, thanks to its incredible versatility. As the name suggests, the jack plane is a jack-of-all-trades and can be tuned for stock removal, flattening and smoothing tasks. As a hybrid woodworker, if you are going to venture

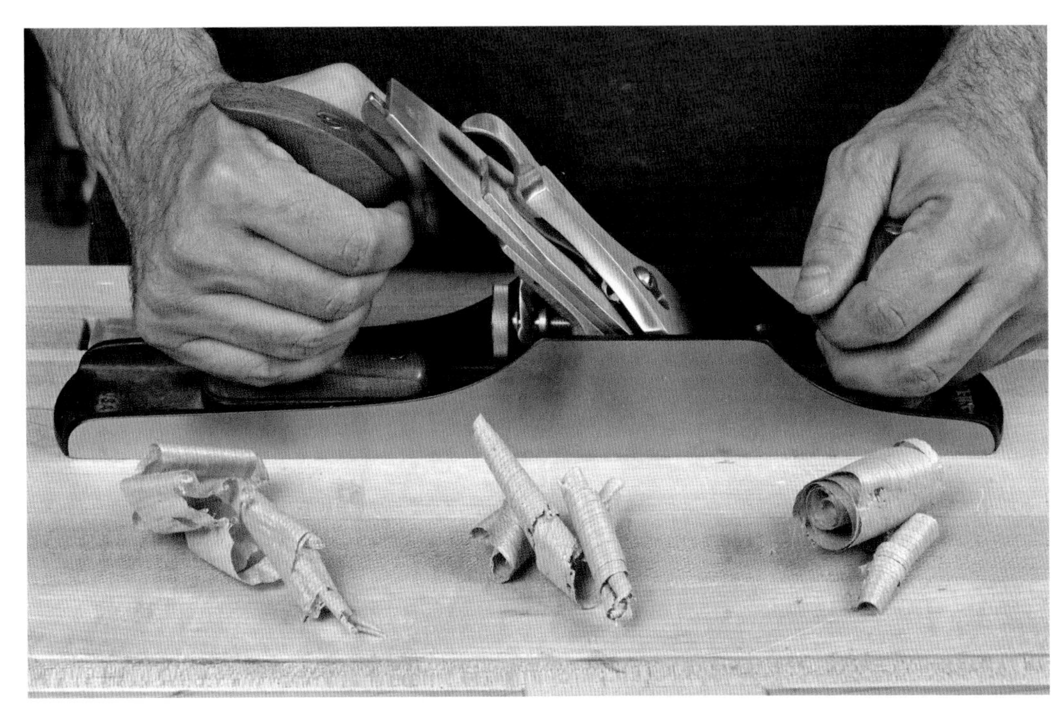

From Smooth to Rough ▪ Just by varying the depth of cut, the jack plane can make a wide range of cuts. Shown here are shavings with thicknesses of .002", .004" and .007".

into the world of bench planes, the jack is a great place to start and could be the only bench plane you need.

How I Use the Jack Plane

With the blade extended for a heavy cut, the jack plane can remove a good amount of stock in a hurry and would certainly come in handy when rough-flattening extra-wide boards (see page 96). Thanks to its relatively long body, anywhere between 14" and 20", the jack plane also can be used to flatten and straighten the faces and edges of boards. And with a finely honed blade set for a super-fine cut, the jack plane can be used as a quick and efficient smoother, making it very easy to smooth panels and fine-tune drawers and doors. These days, we also have low-angle varieties of jack planes that offer even more versatility thanks to the bevel-up design and the ability to swap out blades with various bevel angles.

Bench Planes in My Tool Chest

I currently own a standard No. 5 jack plane and sadly, I almost never use it. My tool collection was assembled over numerous years and like many woodworkers, I have a bad habit of buying first and asking questions later. That's why I already had a smoothing plane and a long No. 7 jointer plane in the tool chest and as a result, the versatility of the jack plane is lost on me. Because I hesitate to part with any tools already in my collection, I reach for the smoother for smoothing tasks and the jointer for flattening tasks, if and when I need them. If I were to do it all again, I would probably choose a low-angle jack plane and divert any unused funds toward other tools that I might be missing.

Form Follows Function ▪ A jack plane falls between a smoothing plane and a jointer plane in size and that says a lot about its functionality as a happy medium between the extremes.

The Workbench

By the strictest definition, a workbench might not be considered a tool. But it is an essential device for any woodworking shop and it deserves to be included in this section. Although many woodworkers refer to their table saw as the "heart of the workshop," hybrid woodworkers and hand-tool woodworkers both see the workbench as truly the center of activity. We may use power tools, but nearly every workpiece ends up secured to the workbench for final processing, even it's just a quick sanding with a powered sander.

In recent years workbench-building has exploded in popularity. Countless books, blog posts, magazine articles, videos and DVDs focus on the topic and one can easily feel overwhelmed by all of the options. Not only that, many folks are quite outspoken about their workbench choices and for some, the workbench itself has become the hobby within the hobby. I don't begrudge this obsession for those who have it, but a little objectivity does go a long way. Like any other tool, we can evaluate the merits of a particular workbench and decide whether or not it will serve our purposes. While I'm sure we all eventually hope to build a dream workbench, reality and practicality dictate that many woodworkers will have to make do with what's available.

Fortunately, a workbench needs only a few basic things to be fully functional. First, it must be sturdy. If the bench moves when nudged, it will most certainly move when running a plane across a board. This is a frustrating side effect of having a lightweight workbench. If your workbench moves, consider adding additional weight. If it has a lower shelf, you can load it up with bags of sand. You might even find a way to secure the bags to the underside of the shelf so you don't lose storage space. My first workbench had a trestle base. It looked beautiful but because it only had two legs, it was light and easy to move. I didn't want to look at sandbags every day so I decided to use the stretcher for clamp storage. With a nice row of heavy-duty F-style clamps hanging on the stretcher, the bench was significantly more stable. Not perfect, but better. However you do it, adding weight will go a long way to making a light bench more user-friendly.

This may sound like nothing more than common sense, but a workbench must be able to hold your workpieces in every orientation the work dictates. Workbench enthusiast and author Christopher Schwarz described it perfectly in the first paragraph of his book

A Woodworker's Best Friend ▪ A good workbench is essential to your success as a woodworker. It doesn't have to be as substantial as this split-top Roubo, but it should hold your work reliably.

Tail Vise in Action ▪ The tail vise holds boards between dogs for easy planing and scraping.

Leg Vise in Action ▪ The leg vise can hold work in a variety of ways, including a vertical orientation for trimming the shoulder of a tenon.

Workbenches: "Every piece of lumber has three kinds of surfaces: edges, faces and ends. A good workbench should be able to hold your lumber so you can easily work on these three kinds of surfaces. Any bench that falls short of this basic requirement will hold you back as your woodworking skills advance." If your bench doesn't have the ability to secure a board for work in all three orientations, you will eventually find it lacking. You certainly can cobble together solutions and use traditional woodworking jigs (plane stops, shooting boards) and tricks to get the most out of a modest bench, even if all you have is a solid-core door and an inexpensive quick-release vise. But you will need to do some research on how to best modify your bench for each workholding task you throw at it. Fortunately, I think this can be done on an as-needed basis and you don't need to have the perfect workbench right out of the gate. But I encourage you to think about the different operations you'll perform at your workbench, which should help guide your decisions about vises and other workholding implements.

A good workbench should also have a reasonably flat top. If the top is flat, you can use it as a reference surface for project assembly and checking the flatness of boards, and of course it will provide a good foundation for every joinery operation you perform on it. That said, I don't feel dead-flat perfection is quite as necessary for the hybrid woodworker as it is for the full-on hand-tool user. The reason is that most hybrid woodworkers mill boards with power tools. By the time the boards get to the workbench, they are already milled flat and square, so it isn't quite as crucial that the bench provide a perfect reference surface. Obviously any major cups, bows and twists should be fixed but please don't spend too much time chasing flatness to the nearest thousandth.

After being a part of the online woodworking community for more than 12 years, I have seen my fair share of butt-ugly workbenches. Yet in spite of their unsightly appearance, these benches provided years of excellent service to their owners. So if your workbench happens to be an ugly duckling, don't give it a second thought. As long as it functions properly, it's probably all you need. But don't let that stop you from building your dream bench when the time comes, because it really is an incredible experience with a reward that you'll appreciate for the rest of your woodworking career.

How I Use the Workbench

The uses for a workbench are nearly infinite, so let's just say that any time a workpiece needs to be secured for any type of hand-held tool work, powered or manual, I use the workbench. I should also add that it supports my cup of coffee quite nicely, with the aid of a coaster of course.

My Workbench

My current workbench is a split-top Roubo design. The bench is based on André Roubo's workbench from plate 11 of his 18th-century treatise on woodworking, *L'Art du Menuisier*. The original design was modified by Benchcrafted to incorporate their amazing hardware. The bench features a 4"-thick top, beefy legs, a strong leg vise, a sliding deadman and a tail vise. I lived with two woefully inadequate workbenches for my first 10 years of woodworking and finally decided to build the bench of my dreams. This bench completely satisfies all of my workholding requirements, whether powered or not.

The most common question I receive about the bench is, "What's the reason for the split top?" There are a few, actually. First is ease of construction. Anyone who has ever made a monolithic slab-top workbench knows how difficult it can be to manage. This top is broken into two slabs, making it easy to maneuver when needed. This made hardware installation easier and also allowed for easier milling because both slabs fit through the planer. The second reason is versatility. Because the gap is there, I can put clamps in the middle of my bench for those oddball projects that need them. When planing drawer sides, I can slip the drawer over one of the top halves and plane away. When not in use, the gap is filled with the gap stop, which also doubles as a planing stop and a tool till. I liked the design so much that I documented the build in a series of 23 videos. You can find out more at http://thewoodwhispererguild.com.

Gap-Stop Tool Holder ■ The gap stop on the split-top Roubo can be used as a tool till while in the middle of a project.

Gap-Stop Plane Stop ■ The gap stop can also be used as a plane stop.

Not Just for Hand Tools ■ A good workbench should also be able to hold the work for power tools.

A Versatile Split ■ The space between the slabs comes in handy when planing assembled drawers.

Should I or Shouldn't I? ■ Many tools fall into a woodworking gray area. Your personal preferences could very well be the only deciding factor.

Hand Tools to Consider

While writing the Must-Have Hand Tools section of this book, I considered not only the hand tools in my own personal collection, but also the wide range available in the marketplace today. The must-haves were easy to identify because they are the tools I reach for on nearly every single project. Most have well-worn handles and blades that have been honed many times over. Deciding which hand tools to avoid was also easy because many of those tools represent expensive and painful lessons in my woodworking history, and we'll discuss those in the next section. Fortunately, many hand tools can be re-sold for as much, if not more, than you paid for them. Thank you hand tool renaissance!

Once I'd sorted out these two extreme categories, there were quite a few tools and tool types that needed a home. These tools are either too infrequently used or too redundant to belong in the must-have category, but because I still own them and occasionally use them, they clearly bring some value to the workshop. Depending on your woodworking budget, style, tastes and habits, you may or may not want to include some of these tools in your kit. And keep in mind, in making these recommendations I'm assuming that the must-have tools are already in your tool kit.

Two Block Planes ■ I like having two block planes, one for rough and messy work (right) and one for fine work.

Planing End Grain ■ A low-angle block plane excels at trimming end grain.

Rough Work ■ Use the inexpensive Stanley block plane for unsavory tasks such as cleaning up glue and epoxy.

Block Plane

The humble block plane is often the first hand tool a new woodworker buys. Many DIYers have a poorly tuned and rarely used block plane sitting in the garage or basement. Given their small stature and low cost, it's no wonder we all seem to snag one, perhaps before we know what to do with it. Although an argument might be made for buying a different hand tool first, you'll likely be purchasing a block plane eventually. These little beauties are equally as handy for hand-tool users as they are for hybrid woodworkers. A well-tuned block plane can not only slice end grain with ease, but it fits easily into a shop apron and can be used for numerous trimming and fitting tasks during the course of a single project. So why is it not in my must-have section? Primarily because of the rabbeting block plane, which I feel is the better initial buy.

Block planes come in two configurations: standard and low-angle. Some also feature an adjustable mouth. The standard-angle block plane is typically milled with a bed angle of 20 degrees. It can take a heavy cut with the grain and is capable of slicing end grain as well. The low-angle block plane typically has a bed angle of 12 degrees and because of the lower angle of approach, excels at slicing end grain while still being capable of fine cuts with the grain. The adjustable mouth (and blade) allows you to dial in the mouth opening to suit the desired-size shaving you want to make. If you want to take a heavy cut, extend the blade and open up the mouth so the thick shavings can pass through. If you want to make a fine cut, retract the blade and close up the mouth so that you just see light passing through. This allows you to make fine shavings while taking light cuts.

How I Use Block Planes

Block planes are handy for many things, so here's a small sampling. When cutting dovetails by hand, I leave the pins and tails slightly proud and use a block plane to flush them to the surface after the glue-up. When easing the edges of a workpiece, a block plane can create a quick chamfer or roundover in seconds, saving me from having to set up my router. Block planes are great for flushing up epoxy-filled knots and repairs. Although I level dowel plugs with a flush-trim saw, I do occasionally reach for the block plane to flush the plug to the surface. After milling, most jointers leave knife marks that appear as a minute washboard pattern. A single stroke or two from a finely honed block plane can smooth the surface to perfection, like a small smoothing plane. When working with plywood, I nearly always attach slightly oversized solid-wood edging to hide the unsightly plywood edges and a block plane makes quick work of leveling the edging to the surface. Of course we can't forget the job block planes absolutely excel at: trimming end grain. Whether it's sizing the length of a tenon or simply cleaning up the exposed end grain of a board, the block plane is perfect for the job.

Block Planes in My Tool Chest

I currently own two block planes: A Lie-Nielsen adjustable-mouth low-angle block plane and an inexpensive Stanley block plane I picked up at a yard sale. Most woodworkers can get away with only a single block plane, but I find having two allows me to dedicate each plane for specific tasks. The Lie-Nielsen is dedicated for wood contact only. You might be thinking, "Well duh?" but block planes are useful for another task; cleaning up glue. For instance, I periodically fill knots with tinted epoxy. After the epoxy dries, I use a block plane to flush the epoxy to the surface of the workpiece. This process is not very kind to a block plane because the epoxy residue gums up the blade and leaves gummy streaks on the sole. So that's where my buddy Stanley comes in, serving as the sacrificial plane that gets used and abused.

Crosscut and Rip Saws

For the most part, crosscut and rip saws are intended for cutting joints. As hybrid woodworkers we already have numerous tools that can help us cut joints so there is rarely a pressing need for a handsaw. The exception here is a small saw for dovetails and other light cutting tasks, as discussed in the previous chapter.

How I Use Handsaws

There have been times when I needed to create a joint at the end of a very large workpiece and one of the more reasonable solutions was to use a handsaw. Maneuvering the large timber onto the table saw would have been difficult and risky. Using the router would have required making a one-time-jig, which would undoubtedly take more time than making the actual cut itself. The dovetail saw would have been ineffective due to its 1½" cutting depth. The only logical option left would be to reach for a handsaw with a larger blade. Let's say the joint in question is a large tenon. A crosscut saw would be ideal for cutting the shoulders and should produce a clean cross-grain cut. The cheeks of the tenon could be cut using a larger saw with a rip configuration,

Joinery By Hand ▪ I don't often cut joints completely by hand, but when I do I rely on a quality handsaw.

because the cut is with the grain. Because I don't often make cuts like this in my work, I saw just outside the layout lines and I recommend you do the same. With enough practice, you can saw closer and closer to the line until the tenons require almost no cleanup. We have plenty of tools for finessing the final fit of the joint so there's no need to risk over-cutting.

Hand Saws in My Tool Chest

I like to be prepared for those important but rare occasions and that's why I still own two handsaws in addition to my dovetail saw: a crosscut carcase saw and a rip-cut tenon saw. The carcase saw features a blade that is about 2¼" deep under the back; that's enough for pretty much any tenon shoulder. There are other saws that feature wider blades for crosscutting, but I have never needed more depth than my carcase saw provides. My tenon saw has a much deeper 3½" blade for cutting the tenon cheeks, which are substantially deeper than the shoulder cuts.

My Three Saws ▪ I have three Western handsaws in my collection for a range of tasks: a large tenon saw, a medium carcase saw and a small dovetail saw.

What About the Ryoba?

I can't depart from this saw discussion without mentioning Japanese ryoba saws. I don't currently own one but it is a saw you might want to consider. As discussed in the Must-Have Hand Tools chapter, the ryoba is quite versatile because it has both crosscut and rip-cut teeth and there is no backbone to restrict cutting depth. If you aren't too picky about tooth configuration, a high-quality ryoba could be the only saw you need. As you can see by my tool collection so far, my saws tend to fall on the Western side of the globe so I haven't felt the need to invest in a ryoba, but it is worth your time to research and consider one for your personal tool kit.

If you already have a dovetail saw and you don't mind making the occasional single-use jig, you could probably make an infinite number of projects for the rest of your life and never require the services of additional saws for crosscutting and ripping. But as you can see from our discussion here, these saws certainly have a purpose and use and might come in handy some day. Frankly, the more you use them the more you are likely to find more uses for them.

Sawing Deep Tenons ▪ Use a deep saw to cut big tenons by hand.

Ryoba ▪ A ryoba, with its two tooth configurations, is a great mutli-tasking saw that makes a thin kerf.

Smooth Plane

Without a doubt, there is nothing quite like the surface left behind by a well-tuned smooth plane. I had the good fortune of working with a $2,000 Brese infill plane at a woodworking show and I swear I saw pixies and unicorns dancing on the wood after each pass. So why do I put smooth planes in the Hand Tools to Consider section? The primary reason is redundancy. While the smooth plane is indeed capable of creating a glass-smooth surface, there are other tools already in our tool kit that can do the job about as well. Cabinet scrapers, card scrapers and yes, even sandpaper, are all capable of producing incredibly smooth surfaces.

While some folks claim to see a visual difference between a finished piece of wood that was handplaned versus one that was scraped and sanded, I don't. Once a film finish has been applied, all surface interaction is with the finish and not with the wood itself. So if the end goal is to apply a film finish it really doesn't matter whether the wood surface was uber-glass-smooth or just glass-smooth to begin with. The difference might be somewhat noticeable when using non-film finishes such as oils. Even then, that difference is going to be minimal at most.

Another factor working against the smooth plane is that it can be difficult to use. Highly figured woods and boards with temperamental grain are notoriously difficult to plane and usually result in major tear-out. The mildest woods can tear out if the smoother is dull or not tuned properly. If you aren't experienced with the tool and its nuances, using it at the finishing stages of a project could yield disastrous results. But make no mistake: In the hands of an experienced craftsperson, the results are practically magical.

How I Use Smooth Planes

In my shop, the smooth plane comes out any time I am working with a wood that planes easily, typically species such as mahogany, cherry and walnut to name a few. In most cases, after I've flattened the boards and cut the joints, the smoothing process begins. Just a couple of passes

Smooth Planes ▪ Smooth planes come in a variety of styles to suit your personal preferences.

Smoothing a Board ▪ A well-tuned smooth plane could be the last thing that touches your work before finish.

Nice Thin Shavings ▪ Smooth planes are capable of taking shavings as thin as .001".

Smoothing Gone Right ▪ When the grain cooperates and the plane is tuned up, a freshly smoothed surface is a thing of beauty.

Smoothing Gone Wrong ▪ Sometimes, the grain sabotages your efforts and tear-out results.

with the smooth plane will save me a significant amount of time over scraping and sanding. But before the plane ever touches the work, I do a few test passes on cut-offs from the same batch of wood. If the test pieces plane well, there's a good chance my critical workpieces will plane well, too. On small workpieces such as aprons, legs and frame parts, where the plane covers the entire width of the piece in one pass, I usually don't do anything further to the surface before finishing. But on wider workpieces that require multiple passes, I like to give the surface one final sanding with #220-grit paper on the random-orbit sander.

Smooth Planes in My Tool Chest

I have used numerous smooth planes throughout the years and I favor the classic Stanley-style metal-bodied varieties for their mass and heft. My current favorite is the versatile Lie-Nielsen low-angle bevel up smoothing plane. Because the blade rests in the plane with the bevel-up, unlike a traditional bench plane where the bevel faces down, its cutting properties can be altered by changing to a blade with a different bevel angle. Pardon the math, but let's examine the options. With a bed angle of only 12 degrees and a typical iron bevel of 25 degrees, the low-angle plane has an effective cutting angle of 37 degrees, which is particularly well suited for smoothing end grain and figured grain, and it does a decent job on standard straight grain. When you encounter tear-out or difficult grain, all you need to do is swap out the blade for one with a steeper bevel angle. A bevel angle of 33 degrees gives a 45-degree cutting angle, also known as "standard pitch". If you have particularly difficult grain and lots of tear-out, increasing the bevel angle to 38 degrees would yield a cutting angle of 50 degrees, or "York pitch". As you can see, a plane like this gives you options.

I have two other smoothing planes in my collection and both are wood-bodied. If you have never used a wood-bodied plane, I highly recommend giving it a shot. I have always been intrigued by wood-bodied planes because they offer a very different user experience. Where the metal-bodied smoother is cold, heavy, industrial and insensitive, a wooden-bodied plane is warm, nimble, organic and tactile. They do take some getting used to, but once mastered, they can be wielded with incredible precision for a gratifying experience.

One of my wooden planes was given to me by my friend Scott Meek of Scott Meek Woodworks. It's a traditional James Krenov-style plane, made by hand, that features a wedge to hold the blade in position. My other wood-bodied plane is a unique offering from Blum Tool Company. The folks at Blum have developed a novel blade-adjusting system with two top knobs for perfect alignment, and that makes it easy to dial in the desired depth of cut. The only disadvantage is that the blades

A Unique Wooden Plane ▪ The wood-bodied Blum smoother has a unique blade-adjustment mechanism.

Fine Adjustment ▪ Using the two knobs on top, the Blum's blade can be adjusted for the perfect setting.

are proprietary and require a special jig for sharpening.

Even when I become proficient with the setup and use of wooden planes, the end result will probably be comparable to what I get from metal-bodied low-angle smoother. But I always leave room in my tool decision matrix for gratification. Some tools are just more fun to use.

Though I don't consider it a necessity for the hybrid woodworker, the smooth plane can be an excellent addition to your woodworking arsenal. As you get more comfortable with a smoother, you likely will find yourself using it more frequently. But because scrapers and sanders can do a similar job with comparable results and with less in the way of specialized knowledge and expertise, the smoothing plane remains firmly planted in the optional category.

Specialty Chisels

The only chisels a hybrid woodworker absolutely needs to survive are standard bench chisels. A basic four- or five-piece set should accomplish most of the chopping, paring and finessing you'll need to do. Because chisel bevels can be ground and honed to different angles for different tasks, it's a good idea to eventually end up with two sets of bench chisels for the ultimate in versatility (see Chisels, page 48). Even with two sets of chisels in the shop, any time your woodworking interests take you into some specialized niche, your standard bench chisels suddenly might not be up to the tasks. Thankfully, there are numerous specialty chisels made to tackle very specific woodworking challenges. I'll discuss a few of the most important types below, but be aware that there are a lot more available these days. There seems to be a chisel dedicated to every possible task we can dream up, though most of them are completely unnecessary for the average woodworker.

Mortise Chisels

Whenever a chisel needs to be hit with great force to do its work, there's a good chance a mortise chisel is the best tool for the job. Whether you're chopping a mortise by hand or simply squaring up a routed one, a mortise chisel is always up to the task. The chisel body

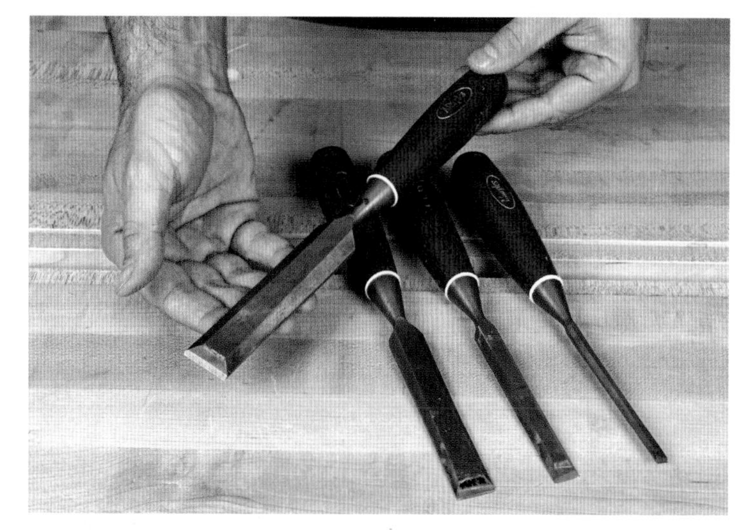

Standard Bench Chisels ▪ Most work can be done using a standard set of bench chisels.

Mortising Chisels ■ Mortising chisels, as expected, excel at cutting mortises. Chopping mortises is tough work requiring a tough chisel.

is very thick with tall parallel sides that help keep the tool square while chopping. The large bevel angle helps the tip of the chisel remain sharp and intact even while taking a substantial beating in dense hardwoods. For heavy work like this, a standard bench chisel might get the job done, but at a certain cost ranging from premature dulling to chipped edges or, in the worst case, a broken tool. While certainly not a requirement for general woodworking, a set of mortise chisels is a worthy addition to your hand-tool collection if you see yourself chopping a lot of mortises.

Dovetail Chisels

Whether making dovetails by hand or using the hybrid band saw method (see page 154), there's always a lot of chisel work. After making the primary saw cuts, we use chisels to establish the shoulder, clean up the pins and tails and finesse the joint for a perfect fit. Because bench chisels are square and dovetails sport angles, it's a spacial challenge to clean up the tight triangular spaces without damaging the pins and tails. In reality, most of us can get by doing the best we can with the smallest chisel in our set. But for those who really want to push the limits with dovetails featuring itty-bitty pins, specialty chisels do have advantages.

Dovetail chisels feature beveled sides that meet the back of the chisel at a sharp point. The body of the chisel is pretty much a low triangle, perfect for cleaning up the triangular pin sockets between tails. Many woodworkers find out the hard way that their standard bench chisels just won't fit into these tight spaces. After mangling their pins and tails they begin searching for a solution in specialty tooling.

If you're strategic about your bench chisel purchase, you may be able to save a few bucks in the long run by eliminating the need for dovetail chisels altogether. There is a variation of the bench chisel known as the bevel-edge chisel. While the bevel usually doesn't meet the back at a sharp point as it does on a dovetail chisel, it does get very close. This varies from

Dovetail Chisels ■ A regular square edge chisel (left) interferes with the angled walls of the dovetail slot. The dovetail chisel (right) can pare right up to the corner.

Tiny Pins ■ If you are a fan of small pins, the dovetail chisel is essential. A normal chisel (above) will quickly dent the sides of the dovetails.

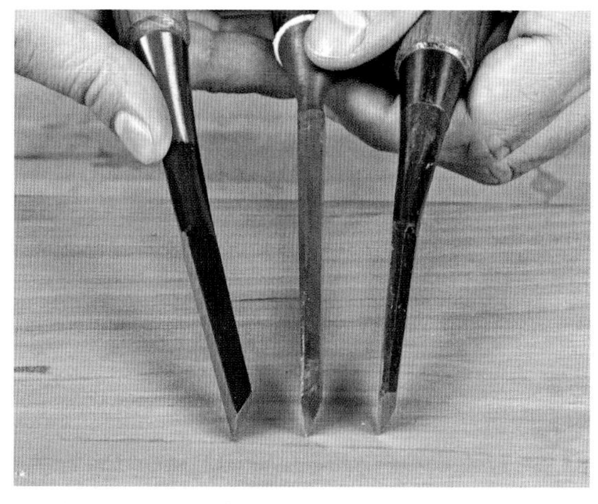

The Edge Makes the Difference ▪ From left to right, a dovetail chisel with a sharply angled edge, a bevel-edge chisel with a small square edge, and a bevel-edge chisel with a larger square edge.

manufacturer to manufacturer so you'll need to do a little homework to find the ones that have the finest edge. I can tell you Lie-Nielsen makes a very nice set of bevel-edge chisels that feature incredibly thin square sides measuring only 20 thousandths of an inch. While not quite the perfect dovetail-cleaning tool, they could very well work for most situations and negate the need for a specialty dovetail chisel.

Fishtail Chisels

The fishtail chisel also is quite useful in making dovetails. This small chisel is unique in that the blade fans out at the tip, which gives it its trademark look. The fishtail chisel excels in cleaning the space between pins of half-blind dovetails. Unlike through-dovetails, where the bottom of the pins and the shoulder can be cleaned up by approaching the work from either side of the board, a half-blind pin board is accessible only from one side. There is a three-walled corner where the base of the pins meets the shoulder. And while a dovetail chisel could clean up part of the joint, it's of no help when chopping the shoulder line right next to the pin. The fishtail chisel, however, easily reaches into the tight corner. Admittedly, this is a difficult purchase to justify as it is really only useful in limited situations. But for folks who plan to cut a lot of half-blind dovetails, this tool could lower your stress level and improve the speed and accuracy of your work.

When it comes to fishtail chisels, if you need one at all, you really don't need multiple sizes. The goal is to clean up a tight corner so once that is done, you can quickly switch back to a regular chisel for the rest of the work. So if you do get a fishtail chisel, get the smallest size available.

Specialty Chisels in My Tool Chest

Because I do a fair bit of dovetailing, I own three different sizes of dovetail chisels and a single fishtail chisel. Although my bench chisels have beveled edges, the square part is about 3/32" thick, far too thick for delicate dovetail cleanup. I absolutely love my bench chisels, but the practical side of me knows I would have been better served by a set of bevel-edge chisels with a much finer edge. Live and learn, my friends.

Collateral Damage ▪ It's easy to see the damage done by this square-edge chisel used to clean up the outside edge of a tail board.

Fishtail Chisels ▪ A fishtail chisel is great for cleaning up half-blind dovetails as well as wider pin sockets.

Maybe Later ■ Moulding planes, scrub planes, long jointers, braces and hand drills, panel and frame saws. You probably don't need them now. You might not need them ever.

Hand Tools to Consider ... Maybe Later

Here's an all-too-common scenario I have observed within the online woodworking community. A new woodworker, let's call him Bob, enters the scene. He has a garage space that he already calls a shop and he's filled it with DIY and/or contractor-grade power tools. He also purchased a block plane from Home Depot but it doesn't work that well so he never uses it. After doing some research, he discovers that many folks are crafting beautiful pieces of furniture the old-fashioned way. He sees pictures of happy hand-tool woodworkers grinning from ear to ear in a sea of gossamer shavings, and he's intrigued. Looking for advice, he leaves a question in a woodworking forum: "I'm looking to get into hand tools. What should I buy first?" Without much in the way of context, well-meaning forum members advise him to buy a good set of bench planes as well as a couple of decent handsaws and of course a good set of chisels. Looking to do his due diligence, Bob purchases a few books and DVDs on hand tools and finds that the people in the forum were right so he breaks out the credit card and buys the handplanes and handsaws he needs. Bob is a smart guy and he already did his homework on sharpening and maintenance, but for some reason, he isn't getting great results. Clearly, this sharpening stuff is a little harder than it looks! The boards he's trying to flatten are anything but flat and the smooth tear-out-free cuts he was promised just aren't happening. Gossamer shavings are replaced with tear-soaked chips and shards. Not to mention Bob's arms are sore.

After taking a few days to clear his head, Bob heads back to the shop hoping for different results. Unfortunately, he is once again left frustrated and bewildered. He looks across the shop at his jointer and planer and sheepishly realizes that the things he's struggling to accomplish with these hand tools, he already knows how to do with his power tools. But the tools have already been purchased and his wife probably wouldn't be too happy to hear that he isn't enjoying using them. Bob is now at a familiar crossroads. One path Bob can take is to fully commit to hand tools. He'll seek out the instruction he needs and he'll practice until he is good enough to produce flat boards and accurate joints. Another path is to go back the way he came, back to the power tools. He'll likely keep those hand tools around with the good intention of learning how to use them later. The third path is the saddest one. Bob is so frustrated that he is prepared to give up the craft all together. In his mind, he clearly isn't cut out for this stuff and decides to try his hand at knitting instead. The thought of any woodworker following the third path is heart-breaking to me. Not that I have anything against knitting, I just don't want to see anyone soured on the craft simply because they purchased the wrong tools in the wrong order.

What could have saved Bob from approaching the crossroads? Starting his journey into hand tools as a hybrid woodworker. The tools outlined in the previous two sections gel perfectly with power tools. They are fairly inexpensive; they tend to be easy to set up and use. While some may interpret my comments here as "anti-hand tool," it is actually quite the opposite. By being set up for early hand-tool success, someone is much more likely to stay interested in the craft and perhaps dive deeper into the world of hand tools as he or she progresses. On the practical side of things, the tools purchased as a hybrid woodworker will still be quite useful if and when they decide to do their best Roy Underhill impression.

As I began outlining the hand tools I felt were must-haves for hybrid woodworkers, it became clear that there also were some tools on the other end of the spectrum, tools that should probably be avoided, at least initially. If a tool creates significant redundancies with respect to power tools or other hand tools, or perhaps it represents a major step backward in efficiency, it quickly found its way into this section: Hand Tools to Consider ... Maybe Later.

Scrub Plane

Many years ago, I began hunting for hand tools to add to my growing collection. Frankly, I was still a "Bob" and I wasn't always sure what I was buying. But I assumed that one day I would figure out what it does and incorporate it into my workflow. Even though I wasn't proficient at hand-tool use, I was well aware of the quality difference between an old flea market Stanley and a brand-new tool from manufacturers like Lie-Nielsen or Lee Valley. My lab-tech salary wouldn't allow new purchases so I decided eBay was my new best friend. I couldn't believe my eyes when I discovered a barely used Lie Nielsen scrub plane up for auction at a full 3 percent below full retail price. I placed my bid then danced a happy jig; I'd won the auction. The plane arrived a week later and I excitedly removed it from the box. Despite seeing pictures online, I was surprised by its shape and couldn't understand why the blade was aggressively curved. I quickly hopped online to do a little more research and realized this tool is intended for rough stock removal. I thought to myself, "Isn't that why I have a jointer and a planer?" Being optimistic, I put the plane on a shelf and decided it would be nice to have just in case I needed it. As the years passed, my confidence in my tool needs grew and the Lie-Nielsen scrub plane was once again posted on eBay for 3 percent off the full retail price.

Scrub Plane ▪ Scrub planes have narrow bodies and no cap iron. The curved blade gouges a shallow furrow in the wood.

Curved Iron ▪ The scrub plane's curved iron, worked diagonally across the grain, peels off a lot of wood in a hurry, but it's only the first step in a lot of planing.

The scrub plane is indeed a rough stock-removal tool and in terms of the flattening process, it's usually the first tool to touch the rough wood. The curved blade allows it to peel away large chips in a hurry, bringing the board to the desired rough thickness. For the hybrid woodworker, this is a 100 percent redundant tool if you already have a jointer and planer in the shop. So unless you plan on milling boards completely by hand, the scrub plane is better left on eBay.

Jointer Plane

Jointer planes are the easiest planes to identify thanks to their incredibly large size. They are also known as "try" planes which I like to think stems from woodworkers telling other people, 'Here, try to lift it!" Coming in at 22" to 30" long, these planes are indeed the largest and heaviest on the market today. Thanks to its long body, the jointer plane excels at making the edges and faces of boards straight and flat. It simply glides over the low spots and trims down the high spots until it eventually takes full shavings. When it does, you can be certain the surface is truly flat.

The jointer plane is the second tool typically used in the manual milling process, right after the scrub plane. Its primary job is to flatten the board along its length and width and it is also used for jointing edges. Because these tasks are easily accomplished with a powered jointer and planer, the jointer plane is completely redundant in the hybrid shop.

In spite of its redundancy, there are times when a jointer plane could be used for a particular purpose such as just flattening boards that are too wide for the jointer (see page 96) or sweetening a machine-jointed edge (page 101). But there are other hand tools in the shop that can do these jobs just as well, notably the jack plane. The jack plane is more versatile and has the added bonus of being cheaper, plus it takes up less space.

I currently own a No. 7 jointer plane and I do occasionally use it for rough-flattening and edge-sweetening. Owning the tool myself, I have to admit that it's heavy handed for me to say one should avoid it completely. It might be more accurate to say you just don't need it. If a jointer plane is ever offered to you for free or you find a really great price on one, go ahead and pick it up. But overall, it's a plane that doesn't need to be in the hybrid woodworker's tool box. If you ever decide to mill your boards completely by hand,

Moulding Planes ▪ Moulding planes do with muscle power what routers do with electricity: shape wooden edges to match their bodies and blades.

the jointer plane will be indispensable. Until then, invest your hard-earned dollars elsewhere.

Moulding Planes

Moulding planes, as the name implies, can be used to create various edge treatments and mouldings for furniture. They come in all different blade configurations and sizes that help you create beads, roundovers and coves. A varied collection of moulding planes can be used in concert to create beautiful complex profiles. Moulding planes are easy to identify by their thin, tall bodies made of wood. Unfortunately, these planes are becoming difficult to find and they tend to be expensive because there arefew makers producing new ones.

Of course, these specialty planes share a purpose with a critical power tool already in our shops: the router. The router can make a complex profile in a single pass in just a matter of seconds. Making an equivalent profile with moulding planes would take significantly more time and multiple plane profiles. I certainly can't begrudge anyone who enjoys the process of creating profiles manually; to each his or her own. But for me, it doesn't make a whole lot of sense.

Mouldings Multi-Tasker ▪ The Stanley No. 45, a Cadillac among moulding planes, plows the groove in a drawer side. Nickers on the body and the skate help make a clean cut.

My advice here is heavily influenced by the fact that my furniture slants somewhat modern. I rarely need complex and unique profiles or mouldings in my work and when I do, my needs are satisfied by specialized router bits or a profile available from a custom moulding shop. I feel it's an overall better investment to have a nice assortment of router bits than to have an entire wall full of moulding planes.

Frame Saws and Panel Saws

There are a number of traditional saws on the market that might catch your attention and tempt you to pull out your wallet. But I encourage you to think carefully about purpose and function and decide if they are truly needed, given your existing power-tool setup. Frame saws, for instance, feature a thin blade held in tension inside a rectangular frame. They can be made in many sizes for various cutting operations but the primary purpose is resawing. The blade is turned perpendicular to the plane of the frame so that as the wood is cut, it passes through the area inside the frame.

You might also come across a variation of the frame saw called the bowsaw, where the blade is on the outside of a wide H-shaped frame. The blade is kept in tension by a twisted cord on the opposite side of the frame. The bowsaw can be used for just about any cutting task including resawing, curve-cutting and joinery.

A panel saw is a large handsaw that can be used to break down boards by crosscutting them to length and ripping them to width. Before I started woodworking, the panel saw was what I thought of as the generic "handsaw."

In the absence of powered alternatives, these saws would be standard equipment for any woodworker. But with tools such as the circular saw, jigsaw, miter saw, table saw and band saw at our disposal, all of our bases are covered and there really isn't a cut we can't make. That makes all three of these tools redundant and unnecessary. While fun to use and equally fun to make (in the case of frame saws), these traditional saws are significantly slower and less efficient than their power-tool counterparts. Furthermore, until the user acquires a good deal of experience with the tool, the cuts are less than spectacular.

I do believe there is one exception for the panel saw. If you don't have a large vehicle for transporting lumber, chances are you find yourself doing board breakdown in the parking lot of your lumber dealer. It's always nice to have a sharp handsaw on hand for cutting lumber

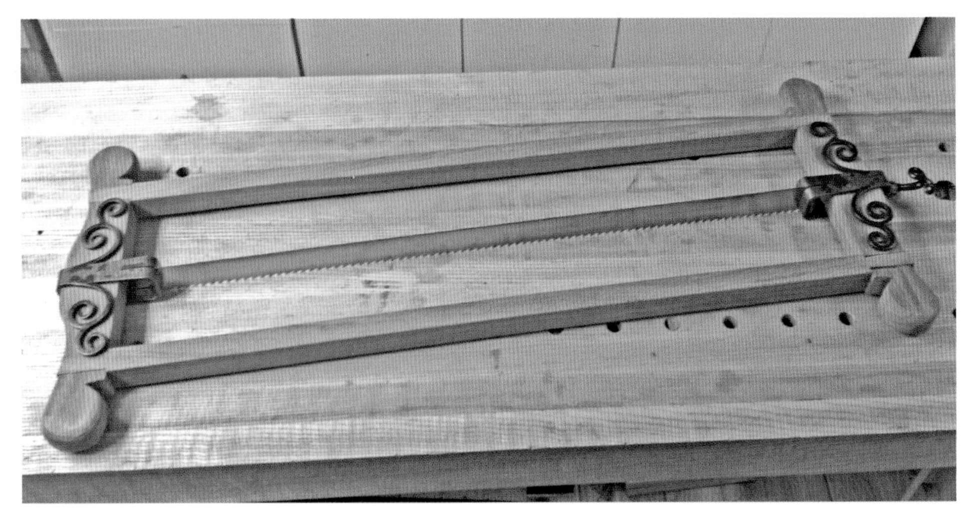

Frame Saw ■ The frame saw is good for ripping and resawing, but not as good as a powered band saw.

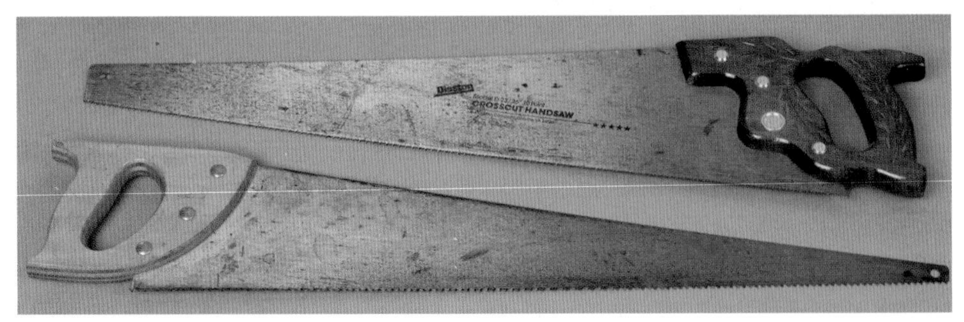

Panel Saws ■ The crosscutting panel saw, top, has 15 teeth per inch, while the rip saw, bottom, has only six.

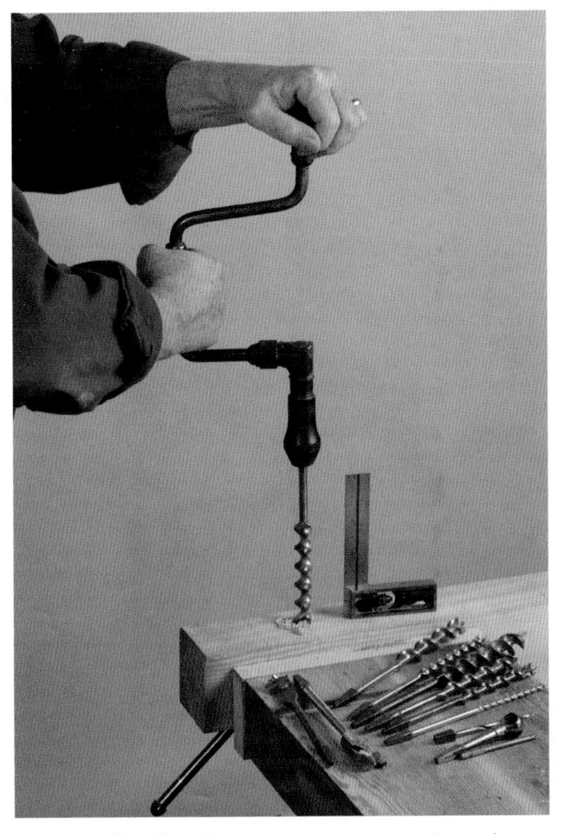

Brace and Bit ▪ The bit brace powers a square-tanged auger bit into the wood. Sighting the square helps keep the hole true.

Hand Drill ▪ The eggbeater-style hand drill can use regular small-size drill bits, but the cordless power drill does everything it could do and lots more.

down to size without drawing a huge amount of attention to yourself. The cut may take longer than it would with a portable power tool but you can keep the saw in your vehicle and never have to worry about dead batteries.

Auger Bits, Brace and Hand Drill

The hand drill is the eggbeater-style tool that some woodworkers use to drill small holes. Its big brother is the brace, typically used for larger holes and auger bits. Both tools rely completely on manual power to do their work and from my perspective, they are completely outdated and unnecessary. Like other tools in this section, they might be fun to use for the sake of nostalgia, but that's about it. Even the cheapest powered drill will do the work of a hand drill or brace in just a fraction of the time, with better results and much less effort.

There is one exception and that's when using auger bits in softwoods. This probably isn't something the average woodworker will need to do all that often but it might come into play when creating ¾" round dog holes in a softwood workbench top. A bit and brace gives you the torque you need at a speed that won't cause burning and does a very effective job of creating the large holes. A powered drill, even at the slowest speed, would likely burn the wood and overheat the bit. So if you find yourself drilling dog holes in a softwood benchtop, borrow a brace from a friend and then return it when you're finished. Most likely, you'll never need it again.

The Hybrid Woodworker Wrangles Bubinga ▪ Without power tools, I wouldn't have been able to mill 75 board feet of 8/4 bubinga, a tough job I would never attempt with hand tools.

Without hand tools, I wouldn't have been able to finesse the platform bed's crisp details and silky-smooth surfaces out of large bubinga glue-ups, nor could I have cut and fit the exacting joints.

The best of both worlds, that's hybrid woodworking. For more on the bed, see page 179.

Techniques Of the Hybrid Woodworker: Machines for Grunt Work, Hand Tools for Finessing

Nearly every power tool available today has its roots in a hand-tool ancestor. Just about every technique performed with a power tool is modeled after a comparable hand-tool technique. Essentially every challenge we face as furniture makers was already solved by clever and resourceful craftsmen centuries ago. I have a great deal of respect for the rich history of woodworking and as a result, there will always be a romantic side of my brain that loves the idea of flattening boards and making joints purely by hand, but the reality is a different story.

At this stage in my woodworking career, I find myself focused primarily on high-quality results, efficiency and personal gratification, in that order. I only use a particular hand tool if it improves my results and doesn't slow me down too much. Because I get a great deal of personal satisfaction out of hand tool use, I always allow for the possibility that going old-school might add a few extra minutes to the process. As long as the results are still high quality, I really don't mind. I am in this craft for the long haul and personal satisfaction cannot be ignored.

Machines Do Grunt Work

While I certainly reserve the right to change my mind over time as I age, mature and become the wise old man I aspire to be, there is one task I can promise you I will never do by hand: mill a board four-square. The first operation discussed in this section is milling and as you'll see, my

Where Power Excels ▪ The milling process is where power tools make the most sense.

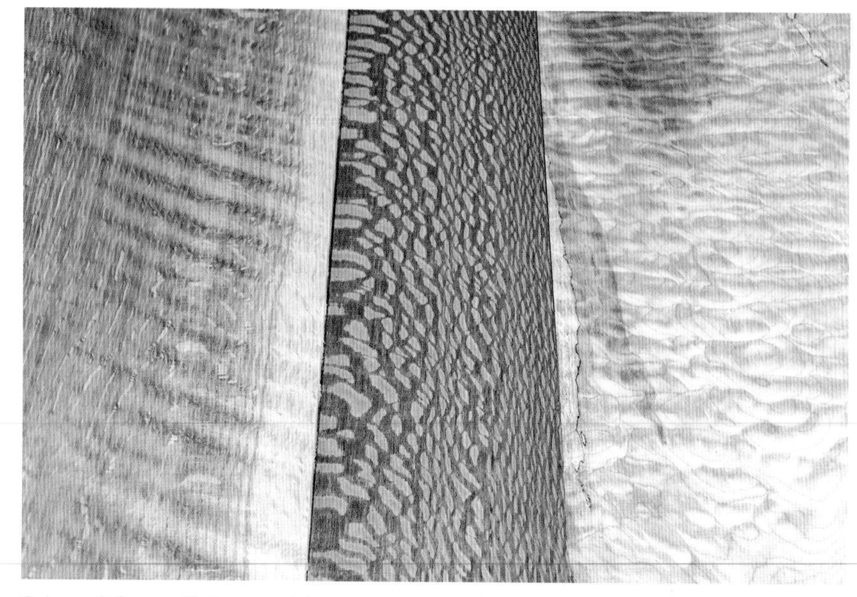

Color and Figure ▪ Nature provides an awesome pallete of colors and grain patterns. Some of these woods present milling challenges.

method is nearly 100 percent focused on power tools. Personally, of all the woodworking tasks required by the craft, milling is beyond a shadow of a doubt the least gratifying. Milling is nothing more than the necessary grunt work that must be done before the fun and exciting work begins. That's exactly why this stuff would have been the job of the apprentice. A hand-tool user might spend all day milling a set of project boards flat and square, which is certainly a great workout, but when it's all said and done they are left with nothing more than a pile of starting material. The power-tool user would have a similar stack of boards ready in a fraction of the time and with greater precision and accuracy. Remember, power tools don't get tired. This is why some of the most devoted hand-tool users I know will still keep a planer in the shop. They know that the best return on their sweat equity is not in flattening boards, but in joinery and finessing.

Beautiful Wood, but Difficult!

Another reason I prefer to mill with machines is the fact that I can power through attractive but difficult woods, allowing me to work with a greater number of species. Have you ever noticed that most hand-tool demonstrations and projects rely on even-grained and softwood species? This is because some species are just easier to process with hand tools. If the wood is dense and highly figured with wild grain, it creates a whole host of unforgiving hurdles for the plane slinger. As a result, hand-tool aficionados tend to avoid the tricky stuff. But in my opinion, some of the nicest hardwoods available are super-dense exotics such as bubinga, wenge, jatoba, and purpleheart, to name just a few. And some of the most beautiful figure you'll ever come across is in the quilted and flame varieties of our own domestic maple. I would be very sad to have these amazing woods taken off of my woodworking palette. Now that's not to say these woods are impossible to work by hand, but they certainly are significantly more difficult and less forgiving. I can see why a dedicated hand-tool user might avoid them whenever possible.

I recently completed a large bed made from 8/4 bubinga that required milling about 75 board feet of lumber. That's a massive quantity of a very dense species and milling that quantity would be exponentially more difficult and time-consuming for a hand-tool user. As the owner of a one-man shop, a project of that size would be impractical, if not impossible, for me to bring to life without the help of my power tools.

Milling Square Stock

Just about everything we do in the woodshop relies upon flat and square lumber as a solid foundation. If we hope to mill the stock ourselves, good milling techniques are absolutely essential to success. You might be surprised to hear that the hybrid woodworking method for milling boards looks exactly like the power-tool method. In fact it's all about power tools, namely the jigsaw, miter saw, jointer, planer and table saw. There is, however, one special case, wide boards (page 96), where I feel a hand tool would be of assistance. Before getting into the details on that, let's review the milling process.

The first step in milling doesn't involve tools at all, at least not the kind you can buy at the store. I'm talking about your eyeballs. Most woodworkers instinctively know to inspect boards for obvious flaws including cracks, knots and excessive warp. At some point in your woodworking career, you'll take this process a step further. Mother Nature gives us an incredible selection of colors and grain patterns to choose from and as beautiful as her work can be, it can also be chaotic. Two boards are never exactly the same and even when cut from the same tree there will be differences that make each board unique. As woodworkers it's our job to organize the chaos, choosing the most appropriate grain and figure for the specific parts of the project. This is easier said than done because in most situations we're presented with nothing more than a pile of boards and a rough sketch of the project we want to build.

Wood Layout

Here's how I deal with it. I focus on the most prominent pieces first, the primary parts of the project: door panels, tabletops, drawer fronts, table legs. It would be nice if we always had an infinite selection of boards to choose from, but the reality is, most of us don't. So by selecting the best grain for the primary parts first, we can at least be sure we're having the most significant impact possible. After the primary parts come the secondary parts: door frames, table aprons, stretchers, shelves. The grain choices for these

Plan Ahead ▪ When milling rough lumber, I like to use chalk to locate parts and highlight flaws to cut away or avoid.

parts are still important, but less so than for our primary parts. When there are only a few boards left to choose from, beggars can't be choosers, but at least the less attractive boards will be used for the less noticeable parts of the finished piece.

Make it Manageable ▪ When cutting large timbers to manageable size, the jigsaw is much easier and safer than the miter saw or a portable circular saw.

Grain Direction Matters ▪ When using the jointer, pay attention to grain direction. Going the wrong way can produce significant tear-out.

Against the Grain ▪ This grain slopes down from right to left, the jointer would be cutting against the grain. Turn the board end for end for a better cut.

Rough Cutting

With the boards assigned their future roles, the next step is to cut them to a manageable size. Using a piece of chalk, I mark each board to indicate which part it corresponds to and measure and mark the rough cutting lines. At this stage accuracy isn't critical but I like to give myself a little buffer, usually an extra inch in length and at least ¼" extra width. Using a jigsaw, I cut the rough boards to length along the chalk lines. I typically do this operation using sawhorses but if the boards are extra heavy, I'll use a circular saw and work right on the floor with the boards resting on rigid foam insulation panels. If the boards require ripping to rough width, I'll use the bandsaw to get the job done quickly and safely. For longer pieces, be sure to use a roller stand on the outfeed side of the cut. When the workpiece is about halfway through, walk around to the outfeed side and pull the workpiece for the duration of the cut. I find this much safer than continually pushing my body and hands toward the unforgiving band-saw blade. You know that's the same tool they use in butcher shops, right? Think about it.

Plane and Joint

With the parts cut to a manageable size, it's time to head to the jointer to plane one surface flat and straight. Before jointing any board, it's important to inspect the grain direction and orient the board for the cleanest cut possible. If you joint the board against the grain, it will not only be difficult to push through the machine but it could also cause significant tear-out. Later in the building process, we'll be smoothing and finessing these boards with hand tools, so it's critical that the surface be as smooth as possible now to save us extra work later.

The easiest way to determine the proper feed direction is to examine the side grain. If the grain lines are sloping down from right to left, that would be considered against the grain because the jointer's cutterhead rotates clockwise. By simply rotating the board 180 degrees with the grain sloping down from left to right, the rotating knives will be cutting with the grain and the result should be a much smoother surface. With one face smooth and flat, we can turn to jointing an edge. The same grain-reading process must occur before jointing the edge except this time, it's the face grain that helps us determine the feed direction. With the jointer fence at 90 degrees to the table, and with the board oriented properly and the clean face against the fence, the edge can be jointed straight, flat and square.

There may be times when no matter how you position the board for edge jointing, you'll wind up going against the grain. This happens when the face grain of the board is traveling at a distinct angle to the edge instead of traveling parallel. To avoid this situation, think about the subsequent edge-jointing step when initially reading the grain. When the face grain has a predominant angle, you should not only be concerned with which direction to feed the board but also with which face would provide the ideal grain direction in the subsequent edge-jointing step. This may sound complicated, but with some practice it will become second nature and within seconds of inspecting a board, you'll immediately have a plan of action in place.

Final Dimensions

Send the board through with the smooth face down to make the remaining rough face smooth and parallel to the first. The cutterhead of the planer is like the jointer so it's important to consider directionality here as well. When inspecting the side grain, if a predominant slope exists, be sure it is running down from the infeed side of the planer to the outfeed side.

Once both sides are flat, flip the board end for end on each additional pass to remove equal amounts of material from both sides as you bring it to final thickness. The board should now be milled to thickness with two flat and parallel faces and one straight and square edge. With the straight edge against the table saw fence, the board can be ripped to final width.

The final step is to trim the boards to length. There are a few ways to accomplish this, including using a miter gauge or crosscut sled at the table saw, but I prefer the miter saw. Outfitted with a high-quality blade, a zero-clearance insert and a sacrificial fence, the miter saw can produce near finish-quality cuts. With a fence system and a stop, it's quick and easy to batch-cut a series of boards to a particular length. If the board's width exceeds the capacity of the miter saw, I'll use the crosscut sled at the table saw to get the job done.

Admittedly, the power-tool milling system does require a significant investment in tooling. But the return on that investment is fast, reliable and repeatable milling. Most important, the efficiency of the system affords me the opportunity to incorporate hand tools in the areas that I find both efficient and enjoyable: joinery and finessing.

Jointer Remedy

If the jointer knives are dull or if the wood is particularly temperamental, the initial jointed edge may not be in very good shape. Assuming there is enough stock to work with, this can be remedied at the table saw by setting the fence about 1⁄16" wider than the part calls for. The rip cut will produce a clean parallel edge and a board that is now 1⁄16" oversized. Readjust the fence for the target width and flip the workpiece over so the fresh-sawn edge runs against the fence. This final cleanup pass will remove any tear-out or mill marks from the original jointed edge.

Small Parts

During the rough-cutting stage, it can be tempting to cut out each and every piece for the sake of organization, but when it comes to small parts, you're better off leaving them un-cut for now. Generally speaking, anything shorter than 12" is what I consider a small part and typically these parts are required in multiples. So for instance, if there is a series of small slats to cut, I'll do the complete chalk layout for reference, but I won't actually cut them out until after the board has been milled to thickness. This not only saves time but it can also be much safer, because planing and jointing boards shorter than 12" is not safe.

The Exception: Wide Boards

As you can see, I use power tools exclusively for all phases of the milling process. But when it comes to milling boards that are a little wider than my jointer, I make an exception. Boards come in varying widths from the lumber mill and it is very common for some to be in the 8" to 12" width range, a size that exceeds our typical 6" and 8" jointers. A common power-tool solution is to rip wide boards in half at the table saw, joint and plane the two narrower pieces and then glue them back together. This can be an effective solution if you truly want narrower boards or when you're trying to flatten a wide cupped board, but if you are looking to retain the full width with no seams, you'll need a different solution. A good hybrid approach involves rough-flattening one face with a handplane, followed by a few passes through the thickness planer. Fortunately, even the smallest planers on the market have a 12" capacity, so if you can effectively flatten one face of the board, you can use the planer to do the remainder of the cleanup.

Inspect the Board ▪ With the board on a flat surface, inspect the perimeter to locate the high and low spots.

Rough Flattening

To begin the process of rough-flattening with a handplane, secure the board at the workbench. You don't need to create a dead-flat surface for this method to work. In fact, if you can just get the outer perimeter of the board into the same plane, it'll pass smoothly through the thicknesser, and you'll accomplish the goal. Low spots are of no concern. Focus on the high spots. To determine where the high spots are, use a combination of straightedges, winding sticks and the benchtop itself. The latter is my favorite method because it's so simple. If the workbench is flat, you can just flip the board over to determine the relative flatness of its face. If the board wobbles or has gaps around the perimeter, it indicates high spots that will need to be removed. Once you determine where the high spots are, you can proceed with the hand-planing operation. Because the goal is to roughly flatten the surface of the board, use a #No. 5 jack or a No. 7 jointer plane set for a fairly aggressive cut. You can use a smoothing plane for this task

Check Your Progress ▪ Use a straightedge to check your progress while planing down the high spots.

Aiming for Flat ▪ When the surface has no major gaps under the straight edge, it is straight. Test in several areas to ensure true flatness.

No gaps = Flat Enough ▪ With the rough-planed side down, there should be no gaps around the perimeter. That's flat enough for our purposes.

as long as you grind a curve into the iron for a more aggressive cut, otherwise you'll be planing for quite a while.

Start by removing the highest spots. If the board has twist, you'll want to begin by attacking the two offending corners. If the board is cupped, begin knocking down the high spots at each edge. I use pencil or chalk to identify the high spots while also gauging my progress. Once you've brought the surface roughly into a single plane, make traversing strokes at an angle across the board. Each stroke should make a complete pass across the width of the board as you work your way down its length. Repeat this process in the other direction and work your way back. Repeat this back-and-forth operation until you start producing uniform shavings with each pass of the plane. Remember, you are really only concerned with the perimeter of the board. Low spots in the middle are of no concern. Every once in a while, flip the board over and observe your progress. Once the board sits stable on the workbench, face down with no big gaps around its perimeter, your bench work is complete and the board should be stable enough for a trip through the planer.

With the rough-flattened face down, the board can be sent through the planer to make the opposing face clean, flat and parallel. The final step is to flip the board and send it through the planer again, completely cleaning up and flattening the original rough-flattened face. The result will be a beautiful full-width board with two flat and parallel faces – and absolutely no glue seams.

Send it Through the Planer ▪ The first pass through the planer, with the rough-planed side down, will clean up the other side.

Flat and Parallel ▪ To clean up the rough-planed side, flip the piece and send it through the planer again. This results in a board with two flat and parallel faces.

Hybrid Methods for Flattening Wide Boards

Dealing with boards that exceed the width of the jointer occurs regularly in most shops. Given the hybrid solution of rough-flattening one side with a bench plane, you might think I made a mistake in the previous tool chapter when I classified both the jack and the jointer planes as something you should only consider buying. While I feel the hybrid flattening solution is effective and efficient, there are other games in town and they involve the tools you very likely already have in your shop: the planer and the router.

Skip-Planing

Skip-planing involves sending the rough boards through the planer without first jointing a face. Take a very light pass on one side followed by a very light pass on the other side.

Skip Planing ▪ If the board is mostly flat to begin with, light passes on each side will yield decent results.

Repeat the process until the board reaches the desired thickness. The idea is to slowly remove the high spots from each side in an effort to eventually end up with something close to a truly flat board. This method can be effective if the board is fairly flat in its rough state. But if the board is bowed or twisted to any significant degree, the planer will not be able to fix the flaws because its pressure-feed rollers flatten the board while it's under the knives. Then the board resumes its original shape after it exits the machine.

Planer Sled

If you think about it, all we need to do to get a bowed or twisted board to pass accurately through the thickness planer is to stop it from moving or wobbling, which is easier said than done. One way to do it is with a plywood sled, some shims, and hot glue. Cut a piece of plywood a few inches larger than the workpiece in each dimension. Place the workpiece on top of the plywood on a flat surface and begin adding shims under the board to balance and stabilize it. Once the board is stable, secure the shims and the board to the sled using hot glue. The entire assembly can then be sent through the planer to clean up one face of the board. Take very light passes during this process until the entire top surface is flat. At that point you can detach the board from the sled, clean up any residual glue, flip the board and send it through the thickness planer to clean up the other face.

Stabilized on the Sled ▪ Hot glue and shims help stabilize the workpiece on a flat plywood sled.

A Pass Through the Planer ▪ The entire assembly goes through the planer to create one flat face.

Router Rails to the Rescue ▪ With a set of parallel rails and a router sled, I can flatten the top of my workbench.

Router Rails

Each of these methods has advantages and disadvantages. One disadvantage they all share is, they are limited by the width of the planer. You can't really use those methods for something as wide as a 24"-wide workbench top. Thankfully, the router-rail method is perfect for such a task. The best analogy I have for this is a CNC machine. Imagine a router with a large straight bit gliding over the work surface in one continuous plane. Any wood sitting in the bit's path is removed and anything sitting below is left untouched. It's a simple concept but does take some work to make it a reality in the woodshop.

The first thing you'll need is a sturdy inflexible carriage for the router. The longer you make the carriage, the more versatile the jig will be on future projects, although anything more than 4' long gets a little difficult to handle. The carriage is constructed from three pieces of 3⁄4" plywood. One piece serves as the base and the other two are the walls. The goal is for this carriage to be as flat as possible and the walls, when glued and screwed to the base, should help keep it flat. The walls also allow the router to fit snugly into the base with only a small amount of wiggle room. The bottom piece of the carriage should have a long slot cut into it that is slightly wider than the router bit you plan to use.

With the carriage constructed, all you need to do is secure the workpiece to a flat surface (hot glue and shims or double-stick tape work well), and create two straight rails for the carriage to ride on. Two jointed and planed 2x4s work well for this, but two identical aluminum extruded bars would be a more permanent solution. The rails should be immobilized on the flat surface on either side of the workpiece with an inch or two to spare, so the router bit doesn't come in contact with the rails. With the carriage resting on top of the rails, try to identify the lowest point on the workpiece and set the bit height so it's touching that spot. Bring the router back to the front of the board, fire it up, and begin making passes from left to right. After each pass, push the carriage forward a bit, and repeat the process until you have routed the entire board clean and flat.

If the board is narrow enough, flip it over and send it through the planer. For larger boards, flip the board over, secure it to the work surface as before, and repeat the routing process. The end result should be a board with two flat faces and a uniform thickness, although the surface will certainly need a little love after routing.

Partial Jointing

This cool trick produces excellent results. Unfortunately, it requires removing the jointer's safety guard, so be on full alert if you decide to try this method. With the jointer set at full-width capacity and the guard removed, begin jointing one face of the board. The board needs a clear path to travel. If your jointer has support surfaces that are outside the cutterhead region where the guard is secured, you may need to loosen the bolts that hold them in place and lower them down as far as they can go. After only a pass or two, you should notice a ridge developing between the jointed section and the unjointed section. When this ridge is detectable along the full length of the board and the jointed section is clean and flat, you can stop jointing.

The next step is to flatten whatever unjointed section remains. If that section is just a strip that's 1" or less in width, I find it easiest to remove it with a block plane and/or a cabinet scraper. Take the board to the workbench and using the flat portion as reference, plane away the high spots until they are flush with the jointed surface. Use a straightedge to check your progress. With the entire surface consistent and flat, you can then pass the workpiece through the planer to clean up the other side.

Now before you forget, go back to the jointer and put that guard back on.

The Hybrid Twist

Twist is the most challenging defect to correct. The board has two high corners and two low corners. If you try to joint the board, it can be incredibly difficult to focus pressure in a way that actually removes the twist, instead of making it worse. The thickness planer isn't much help either because it tends to follow the shape of the board. Without a flat reference surface, the thickness planer simply makes a thinner version of the same flawed board. There are a few alternative power-tool options available, such as router rails and a planer sled, but they require quite a bit of setup and jig/fixture-making, which may be more work than you want to put into a single twisted board. This is why I find the hybrid solution to be attractive. It only takes a few minutes to remove the twist with a good sharp handplane and when one face is mostly flat, it's safe to pass the board through the planer.

Chalk it Up ▪ Using chalk, mark the high corners that need to be planed down.

Work at an Angle ▪ Push the plane at an angle across the board, focusing on the high corners.

Built-In Straightedge ▪ The plane itself works as an effective straightedge to evaluate the flatness of the board.

Back-Lighting Sheds Light ▪ With a strong backlight, it's clear what areas need more work. The board in this photo is flat enough for most purposes.

Panel Glue-Ups

The first time I learned that two boards could be joined together in seamless unity, I was amazed. It is something experienced woodworkers take for granted but it truly is a special thing. Making narrow boards look like wider boards is central to the furniture-making process. With careful grain selection and preparation, the seam or glue line between the boards can be made all but invisible and the onlooker will have no idea the piece is made up of multiple pieces of wood. Because we almost never have boards wide enough to make a full tabletop or a cabinet side, joining boards is critical, and it's a task that certainly can benefit from the hybrid approach.

I do rely on power tools for the initial milling of the boards that are destined to become one. In many cases, you can simply take two jointed boards and glue them together edge to edge. But what many folks don't realize is that the jointed edges will sport milling marks and are far from perfect. With the exception of finely tuned jointers with ultra-sharp helical cutterheads, jointers leave a minute washboard pattern on the planed surface. When you glue two washboard edges together, the result is a too-wide gap between the boards and a more obvious glue joint in the product. When it comes to panel glue-ups, you want to be as close to perfect as possible, in hopes of creating a nearly invisible joint. Fortunately, all you need is about 30 seconds and a block plane, a smoothing plane or a cabinet scraper to make a simple but significant improvement.

Sweeten the Edges

After jointing the edge of the workpiece, secure the board to the workbench. Sweeten the edge ever so slightly, doing nothing more than knocking down the peaks of the washboard pattern. If the wood is a very cooperative species and you don't anticipate tear-out issues, use a finely-honed block plane or smoothing plane to do the work. Try to take the tiniest shaving possible with the blade barely protruding from the sole and with the mouth, if it's adjustable, closed up so only a small strip of light shines

The Width We Need ▪ By milling the parts with care, we can glue up multiple boards to get the appearance of a single wide board.

Not as Smooth as You Think ▪ Most jointers leave a washboard surface that only becomes visible when rubbed with chalk. This ribbed surface can make a less-than-perfect glue joint.

A Mini Smoother ▪ A finely honed block plane, set for a light cut, quickly removes the ridges left by the jointer and leaves a smooth surface.

Scrapers Work Too ▪ With only a little amount of material to be removed, the cabinet scraper does a fine job of knocking down the high spots on these walnut and maple (at right) edges.

Scrapers for Difficult Grain ▪ A block plane could make a mess of temperamental or highly figured grain. The cabinet scraper is more forgiving.

Invisible Joint ▪ These two boards are just butted together with no glue and no clamps. When the edges are perfect, it isn't difficult to get a good joint.

through. Although the edge has a washboard pattern, it is a consistent pattern and the edge is still considered flat, so it's easy to balance the plane securely during this operation. After two or three passes, the surface should be smooth, straight and free of irregularities.

If the wood is more temperamental, I'll do this process with a cabinet scraper. The scraper is much more forgiving on difficult woods and thanks to its wide sole, knocks down the peaks just like the block plane and smoothing plane. And while I'm sure someone might have success using a card scraper for this operation, I don't recommend it. The card scraper, without a sole to guide the cut, follows the hills and valleys on the surface and may also create an out-of-square edge.

If you are well versed in bench planes, you might be wondering why I don't recommend using a jointer plane for this edge-smoothing task. After all, jointing an edge is the job the long jointer plane excels at. First, let me say that you certainly can use it for this operation if you choose to. Set the blade for a fine cut and go to work. Just keep in mind that this smoothing process is not really intended to be a jointing operation. The surface is already machine-planed square and straight, and in the world of hand planes, we are now moving into a smoothing step. There is no longer a need for the long plane body of a No. 7 or No. 8 jointer plane. That's why I prefer to use something with a much shorter body that will follow the path of the surface already established: the block plane, cabinet scraper or smoothing plane. As long as you remove a uniform amount of material along the length of the edge, you will maintain the straightness established by the powered jointer.

Whichever tools you use to accomplish the task, you'll be rewarded with a much smoother edge that will mate perfectly with the adjoining edge. The gap between the boards will be negligible and the resulting glue line will be nearly invisible. If you match the wood figure carefully, the trained eye might have trouble spotting the seam.

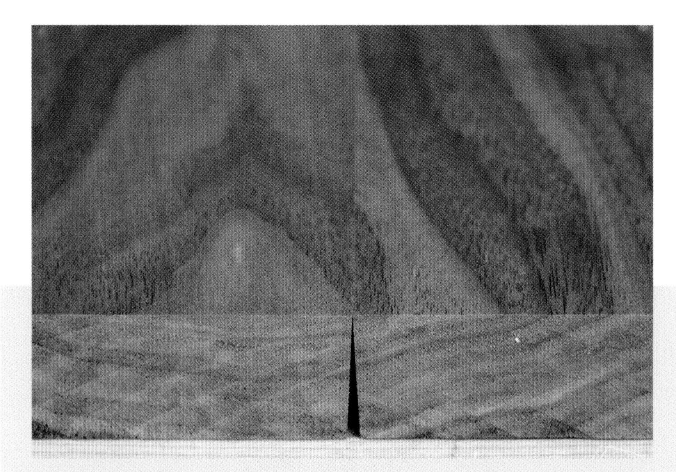

Out-of-Square Edges ▪ If the jointer fence isn't at 90 degrees to the table, both boards will have a small error that results in a big gap when they're butted up edge to edge.

Work Two at Once ▪ Running a plane over two edges at the same time cancels out the off-90-degree angle.

Perfect Edge Joints

Speaking of square edges, here's an additional trick to help edge joints mate perfectly. Unfortunately, some jointers will leave an edge that is just shy of a perfect 90 degrees. This could be due to poor setup or more than likely, a warped fence. Either way, you can orient your boards so that the two off-square edges offset one another, essentially nullifying the error. To accomplish this, I sandwich two boards together with the two adjoining edges on the same side of the sandwich and proceed with the sweetening process. By planing or scraping both edges at the same time, we can be confident not only that the high points are removed, but also that the two edges feature the same angle. When it's time to do the glue-up, unfold the boards like the pages of a book, keeping the two jointed edges together. This orientation allows the two angles to offset one another and the resulting joint should be perfect. Even if you know your jointer produces perfect 90-degree edges, you can save significant time by sweetening two edges at once.

Perfectly Imperfect ▪ Matching the two off-90-degree angles nicely closes the joint. The angle is exaggerated in this photo.

Flattening a Panel Glue-Up

Once a panel has been glued up, regardless of the quality of the edges, there will almost always be some cleanup to do wherever two edges meet. The primary reason for this is the properties of glue. When two wet glue surfaces meet, they can easily slip past one another. As you tighten the clamps, two perfectly aligned boards may wind up slightly offset. If you've milled the boards to the same thickness and paid careful attention to alignment during the glue-up, this discrepancy can be minimized. I like to stack the cards in my favor by using alignment helpers such as biscuits, Dominos or even dowels. Clamping long cauls across the joint is always a good option, too. With this extra assistance, I can apply the glue and clamp the boards together without having to stress about perfect alignment.

Even after taking all possible precautions, there is always some surface leveling to do in order to achieve that seamless look and feel we all strive for. If you are lucky enough to have a wide-belt sander or a drum sander, the surfaces can be leveled in just a few minutes with clean results. And while you might be tempted to have at it with a portable belt sander, it's dusty, nasty work that rarely results in a clean surface.

But if the board is too wide for the drum sander or you don't have a drum sander at all, you can use hand tools to get the job done.

You can level the surface with any of several tools, including a block plane, a cabinet

Glue it Up ▪ With glue on both mating edges, clamps pull the joint together and squeeze out a little bead of glue.

Cauls Help Alignment ▪ Glue may cause the two mating edges to slip, but wooden cauls clamped top and bottom keep the glue-up flat.

Domino for Perfect Alignment ▪ The Festool Domino creates small loose-tenon joints that are perfect for aligning panels.

Quick and Accurate ▪ The precisely located mortises and pre-made Domino tenons make panel glue-ups a no-brainer.

scraper, a card scraper or a smoothing plane. If the discrepancy is greater than 1⁄16" or so, I recommend using the block plane, set for a fairly aggressive cut, to level the area around the joint. You might also consider using a larger-bodied plane such as a smooth or jack plane to make quick work of the initial leveling task. Fortunately, with proper milling and careful attention to alignment during the glue-up, most discrepancies should be less than 1⁄16" and won't require the services of large, aggressive planes. In fact, I do the vast majority of my panel smoothing with the cabinet scraper.

Cabinet Scraper

With the scraper set for an aggressive cut and starting on the lower of the two boards, push the tool diagonally across the joint. Cutting at a slight angle reduces chatter and makes for an easier cut. Initially, I focus only on the area directly alongside the glue joint. Once the joint itself feels flush, I begin making longer passes, extending the path of the tool across the entire surface. Although the joint may feel flush, the surface as a whole will no longer be flat. By extending the path of the tool, we can gracefully and gradually level and blend the surfaces so they are as flat as possible.

When it's all said and done, the joint usually is impossible to locate using the most sensitive tool I own: my finger. With the initial flattening complete, I set the cabinet scraper for a very light cut and take several passes over the entire panel. In the same way the jointer leaves mill marks on the edges of a board, the thickness planer does the same on its faces. So it's a good idea to take a quick sweep over the entire panel to remove any remaining mill marks. Because my cabinet scrapers have a really aggressive hook on them, I like to use a finely tuned card scraper for the last few passes. Each tool does tend to leave marks of its own, and much like progressing through the grits of sandpaper, I like to use progressively finer and finer tools to smooth the surface.

Some woodworkers consider a freshly scraped surface ready for finish, but I don't. Even the card scraper can leave behind small marks that

Cabinet Scraper For Flattening ■ After glue-up, the No. 80 cabinet scraper removes material quickly without the risk of tear-out, and its wide sole helps keep the panel flat.

Card Scraper Works Too ■ A standard card scraper may also be used to fine-tune a panel glue-up.

Finish with Sanding ■ A #220-grit sanding using the random-orbit sander makes the panel ready for finishing.

may not show themselves until the finish is applied, so I recommend finishing up with a light sanding using #220-grit paper and a random-orbit sander. I don't feel the need to eliminate sanding altogether because it is still the best ways to achieve a smooth and consistent surface. Using the hybrid methodology, however, we can dramatically reduce our dependency on sandpaper, which is good for our lungs (less dust in the air) and our bank accounts (less sandpaper to purchase).

A Good Soak ■ Use a damp rag to drip water directly onto the dinged area.

Steam Treatment ■ Using a hot iron, press a wet rag into the ding and let the steam penetrate the wood grain. The damp wood will swell and expand slightly.

Flatten When Dry ■ After the water evaporates, clean up the raised area using a scraper.

Bye Bye Ding ■ If all goes well, the swelling wood fibers will fill the ding completely, making an invisible repair.

Ding!

Inevitably, your freshly smoothed boards will receive an unfortunate ding or scratch from a wayward tool or perhaps from contact with another board. Fortunately, many dings and scratches are easily removed by steaming. Using a wet cloth and an iron, steam the defect in short bursts. As the wood fibers absorb the moisture, they will expand and fill the void. If the dent is shallow, it is very likely that all you'll need to do is give the surface a light sanding and it will look as if nothing ever happened. For deeper injuries, and those where wood was actually removed and not just compressed, there's a little more work involved. After steaming, I begin working the area with a card or cabinet scraper. Even though the flaw only exists in a small isolated area, you'll want to be careful to fan out with each pass and extend the material removal beyond the flaw. This will help keep the surface looking and feeling flat and prevent creating a noticeable dip.

What About a Smooth Plane?

If you are so inclined, you can always perform the final smoothing steps using a smoothing plane. There truly is nothing quite like the feel of a surface that was just hit with a finely tuned smoother, but numerous things have to go right for this magic to happen. First, the wood itself must be user-friendly and not prone to tearing out. We are talking about finish preparation, so at this stage we simply can't afford to risk tear-out. I recommend practicing with scrap material to confirm that your stock responds well to the smoothing plane's advances. A second issue may arise when you realize your panel is not truly flat. When making a wide surface such as a tabletop, the wood will inevitably move to some degree. Wood just doesn't stay dead-flat for long. Additionally, if the glue joints were uneven and some flattening was required, it's very likely that the surface will have numerous subtle peaks and valleys. These imperfections are usually imperceptible to the human eye but a smoothing plane will certainly let you know they exist, showing themselves as "holidays" in the otherwise glass-smooth surface.

For a smoothing plane to work properly, it must be used on a surface with no valleys longer than the length of the plane's own sole. If during the leveling process you created some large but visually undetectable valleys, you will have significant trouble getting full-length continuous shavings with a smoothing plane. The third thing to be concerned about is whether the cut quality truly is finish-ready after smoothing. On large surfaces where multiple passes are needed, a smoothing plane, much like a card scraper, will leave at least some evidence of its passing. As a result, most folks find themselves still doing one final sanding to make sure the surface has a smooth and consistent look and feel. On smaller panels and smaller furniture parts such as aprons and legs, I use the smoothing plane for final processing because the blade typically covers the workpiece in one single pass. On larger surfaces, I pass on the smoothing plane and instead reach for my trusty random-orbit sander.

Smooth Plane for Smoothing ▪ The smooth plane also makes short work of final smoothing tasks.

Smoothing Gone Right ▪ If your plane is tuned up and wood cooperates, a smoothing plane can leave a surface that begs to be touched.

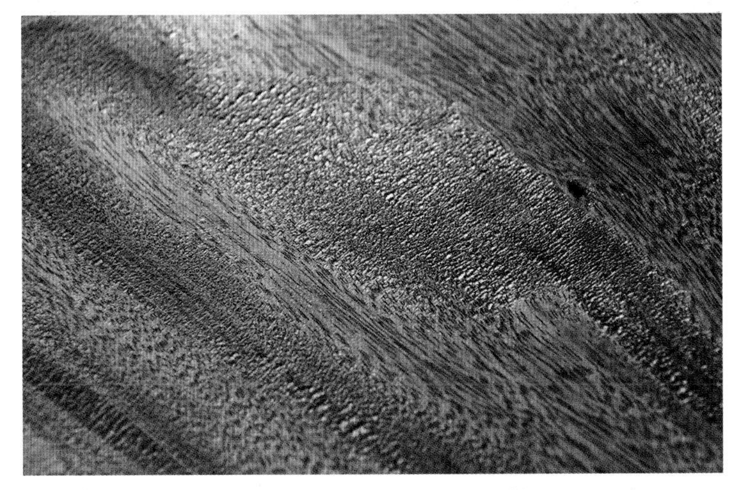

Smoothing Gone Wrong ▪ If the plane iron is dull or if the wood isn't cooperating, the smoothing plane is liable to tear up the wood grain.

Getting Groovy ■ These three grooves have specific names determined by their location and grain orientation. On the left edge is a rabbet. Perpendicular to the grain is a dado. Parallel to the grain is a groove.

Dados, Rabbets and Grooves

Dados, rabbets and grooves are essential joints in cabinet construction as well as in other areas of woodworking. While all three could be called "grooves" by the lay person, woodworkers give them three distinct names based on their location and orientation.

A dado is a groove cut into a board perpendicular to the grain. An example of this would be a dado cut into a cabinet side to receive a shelf. The groove is perpendicular to the grain or cross grain as it runs from one edge of the panel to the other.

A groove is similar to a dado, except it's oriented with the grain instead of cross grain. We typically see grooves in drawer sides to capture the drawer bottom, and in door frames to capture the door panel.

A rabbet is a groove cut right at the edge of the workpiece so it's open on one side. We often cut rabbets around the back perimeter of a cabinet, so the back panel can drop in place. These joints might be found on any project.

Making dados, rabbets and grooves with hand tools is a lot of work. Layout, sawing, chopping and planing a few grooves might be fun on a small project, but for anything larger than a little jewelry box, I prefer power tools. While tools such as the router and table saw may help us make these joints faster and more efficiently, the results aren't always perfect. Thankfully, the hybrid woodworker has hand tools to help dial in the fit of every dado, rabbet and groove.

When cutting dados, rabbets and grooves, I use the table saw whenever possible. The setup is quick and the results are predictable. But some pieces are awkward or unsafe on the table saw and for those, the router is my weapon of choice. Tablesaw setup will vary depending on the orientation of the workpiece. For narrow grooves, use a ripping blade or a blade with a square grind. Either will leave a flat-bottomed groove with little to no ridges. For anything more than 1⁄8" wide, use the dado stack. If the joint runs in the long dimension of the workpiece, dados and grooves can be guided by the standard table saw fence. But in the case of rabbets, where the blade needs to cut at the very edge of the workpiece, add a sacrificial fence that allows a small part of the saw blade to nest within it.

When the cut runs in the same direction as the short dimension of the workpiece, which is usually the cross-grain direction as well, you need to take extra precautions. I use either the miter gauge or a crosscut sled to guide the workpiece. If the operation feels awkward or even remotely unsafe, I depart from the table saw and use a router with a straight bit and a guide instead.

The Problem

While the table saw does a fine job of creating grooves quickly and efficiently, one side effect can haunt the woodworker: inconsistent groove depth. Two physical factors work against us when cutting grooves at the table saw. First is the workpiece itself. Many times, plywood and other large panels are not perfectly flat, although they are still usable in the project. So if the piece bows slightly as it's pushed across the blade, and if you don't apply substantial downward pressure, the depth of the groove may not be consistent. Second, the dado stack. Because the dado blades have to remove a lot of material at once, the workpiece tends to lift ever so slightly. Even if you're using a dead-flat piece of plywood, it is likely that the workpiece will lift at some point during the cut, causing a high spot in the groove. As inefficient as it may be, I often run the workpieces through the saw twice at the same setting in an attempt to alleviate this problem, though it does introduce the possibility of a new problem: accidentally widening the groove, resulting in a loose fit.

Here's why a high spot in the groove is such a problem. Suppose you are building a bookcase with a rabbet at the top, a dado near the bottom, a dado for a fixed shelf in the middle and a rabbet around the back for the back panel. Blissfully unaware of this uneven depth issue, you cut the joints and assemble the bookcase. Once the glue dries, you may discover that the sides seem off somehow. A long straightedge on the side reveals that the middle of the case

Rabbets at the Table Saw ▪ By embedding the dado blade in a sacrificial fence, rabbets can be cut quickly and easily at the table saw.

Guided Router Cut ▪ The router and edge guide makes a powerful combination. With careful setup, you can rout dados and grooves on workpieces that might be difficult or impossible to manage at the table saw.

bulges outward. How could this happen? Inconsistent groove depth. The workpiece must have lifted slightly while you were cutting the fixed-shelf dado or the back rabbet. So it is crucial to inspect the grooves thoroughly and fix them as needed. And this really underscores the general importance of the dry-fit process because it's always better to discover these issues before you apply any glue to the joints.

The Hybrid Solution

The hybrid woodworker can avoid inconsistent grooves by using the trusty router plane outfitted with the widest square blade that fits into the groove. With the workpiece secured flat on the workbench, set the router plane to the desired depth and pass its blade through the groove channel. Anything higher than the blade will be pared away. If the groove is perfect, the plane will slide from one end to the other without making contact.

Dados and grooves are easy to fix with the router plane because the surrounding workpiece offers perfect support on both sides of the tool. It's one of the few woodworking operations where you can check your brain at the door and still get good results. For rabbets, there will only be support on one side of the plane so it's important to apply extra downward pressure on the supporting side or use an extended auxiliary base (see page 130). If you have more than one rabbet to clean, you can always butt two rabbeted workpieces together edge-to-edge and finesse them both at the same time, with the router plane fully supported on both sides.

If there's a lot of material to remove, try backing off the plane blade a bit and proceeding in small steps. Too big a bite and the plane will be difficult to push through the wood, leading to tear-out. On plywood, the internal plies are often made with substandard wood and the grain direction changes with each layer, so

Dado Cleanup ■ A router plane ensures all dados and grooves are cut to a consistent depth.

after planing plywood, dados aren't exactly the prettiest thing to look at. But if the depth is accurate, who cares? No one will ever see it after assembly.

For narrow grooves, the Veritas router plane has blades available all the way down to ⅛". Although I don't own one, there are smaller router planes on the market with narrow blades. Veritas also makes a small plow plane that has interchangeable blades down to ⅛" wide. This specialty plane has a fence and a depth stop and can be quite handy for cleaning up narrow grooves.

What about router plane alternatives? Although the router plane excels at cleaning up dados, rabbets and grooves, some people will choose to use a shoulder plane or rabbeting block plane for some of these tasks. If the plane fits within the joint's walls, there's certainly no reason it can't be used. The shoulder plane and rabbeting block plane have no depth setting so the only thing stopping you from removing too much or too little material is frequently checking your progress by measuring the depth. The router plane, with its fixed depth setting, provides a vastly superior user experience. You just set it to the appropriate depth and go.

Stopped Dados and Grooves

Sometimes, instead of going all the way through the workpiece from one end or edge to the other, the groove stops short. This variation conceals the joint from the front of the case. It's also used in a drawer so the groove for the bottom doesn't show itself through an elegant set of dovetails. Just like through-dados and grooves, hand tools are tedious and power tools can result in inconsistent groove depths, so we'll once again rely on power tools for the grunt work and hand tools for finessing. But there is one additional issue at play when it comes to a stopped dado or groove, and that's the accuracy of the stop point itself. The stop needs to be precisely located and square, and no power tool in the world can make that happen. The only way to do the work accurately is with a good sharp chisel. And because the chisel gives you a great deal

Work Smarter ■ Cleaning up rabbets with a router plane can be tricky unless you butt two rabbeted boards together.

Yes, Plywood Too ■ A router plane is not just for solid wood. I use mine frequently when working with sheet goods such as this piece of plywood.

Helpful Lines ▪ The long mark on the sacrificial fence shows exactly where the blade starts cutting.

Match up the Lines ▪ When the line on the workpiece touches the line on the fence, you know you've gone far enough.

The Ramp ▪ The circular saw blade leaves a small ramp at the end of the cut. It needs to be removed.

Cleaning Up the Dado ▪ Use a chisel to remove the ramp and complete the stopped dado.

of control, the crucial stop point can be located perfectly.

Using the table saw to hog out a stopped dado or groove can be a tricky business because the cut doesn't go all the way through the workpiece. To make the cut, we need to mark the exact point the blade starts cutting, either on the saw's insert or fence, then we need a second mark on the workpiece to show where the cut needs to stop. Then we push the workpiece over the blade until the two marks match up and turn the saw off while holding the workpiece in place until the blade stops, or else simply draw the workpiece back to the start of the cut. I don't like this method because it's not safe. I'm not a fan of holding any workpiece still over a spinning blade, ever. I am also not too thrilled about pulling the workpiece back toward me while the blade is still spinning because this is a great way to create wooden rockets. Additionally, this method results in a long curved ramp of material leading from the bottom of the groove up to the stop point.

All of that extra material will need to be pared away with a chisel to achieve a flat, consistent bottom and a square end. Because of the associated safety concerns and the additional chisel work, I prefer the router for cutting stopped dados and grooves.

The big advantage to routing is the fact that you can get very close to the stop point with the groove at full depth. Once the groove has been routed right up to the stop-point pencil line, I make sure the groove is at a consistent depth using the router plane. Fortunately, unlike a table-sawn groove, a routed groove is almost always at a full consistent depth, thanks to gravity and the heft of the router itself. The only thing left to do is square off the end of the groove using a chisel.

Stopped Dado With a Router ▪ The router also can make a stopped dado, leaving a small amount of material to square up at the end of the cut.

Consistent Depth ▪ A router plane ensures the dado is a consistent depth, although unlike with the table saw the results from a router should be near perfect without planing.

Chiseling Square Ends

First, establish the perimeter of the stopped groove. Rest the back of the chisel against the groove wall with part of the chisel over the uncut surface and roll the chisel forward to slice the wood fibers up to the pencil line. Because you're using the groove wall itself for reference, the cutline should be straight and perfectly aligned with the existing groove wall. Repeat the process on the other side. Next, establish the end of the groove by placing the chisel tip right on the pencil line and give it a few taps with a mallet.

Continued on next page

Extend the Walls ▪ Extend the walls of the stopped dado, using the sawn walls to guide the chisel.

Mark the End ▪ Using light taps, establish the end of the dado right in front of the layout line.

Don't hammer too hard when initially establishing the line because the chisel's bevel is still in contact with the wood in front of it. The wedging action can compress the fibers and push the chisel farther back than you intended. With the outer perimeter scribed, hold the chisel horizontally and carefully pare away the top layer of material. Repeat the entire process by re-defining the outline and then paring away the waste again. For most standard grooves, three to four rounds of paring is about all you'll need. With the walls established and the corners nice and square, the last step is to make sure the groove floor is smooth and consistent. A router plane can be used for this or simply scrape the bottom with a chisel, bevel down.

Pare Away ▪ Carefully pare away the top layer of wood. Outline again and pare again.

Final Depth ▪ With most of the waste removed, chop down to full depth right on the layout lines.

Clean the Bottom ▪ The bottom of the dado will likely need some work.

Bevel Down for Best Results ▪ With the bevel of the chisel down, pare away any proud wood and carefully work back to the end of the dado.

The Mortise-and-Tenon Joint

The mortise-and-tenon joint is fundamental in woodworking and all woodworkers see it as a mark of quality construction. Throughout the years, woodworkers have found countless ways to make both the mortise and the tenon, and most of them are completely valid. Remember, there's always more than one way to get a woodworking task done, and after much research and experimentation of my own, I have settled on a system that gives me the accuracy I need, the speed I want and the gratification I crave.

Anyone who has tried to construct a perfect-fitting mortise-and-tenon joint knows just how many different surfaces are at play. If any one of them is just a little too big or too small, the joint won't fit properly and the structural integrity will suffer. So it's vital that we use a system that allows us to fine-tune the joint quickly, accurately, and with little chance of removing too much wood. That last point is critical because power tools don't often allow us fine control. I like to joke with people by telling them I am capable of adjusting my tablesaw blade height in .001" increments, with only +/- ⅛" error. In other words, a slight adjustment at the table saw can easily take a tenon from a bit too snug to sloppy loose.

Thanks to the help of a few hand tools, the hybrid method gives us the ability to finesse the fit of the tenon with a great deal of accuracy and control. The mortise is cut completely with a router, providing the fixed reference for the tenon. The tenon is roughed out at the table saw, then finessed at the workbench using planes and chisels. Because hand tools have the ability to remove shavings of just a few thousandths in thickness, we have a great deal of control over the final fit. Best of all, should we go just a little too far by making one or two extra passes, the joint will still be within the range of an acceptable fit.

The Mortise

A mortise is a recess cut into a piece of wood. In the case of a mortise-and-tenon joint, the recess is cut to the exact dimension that will house the tenon on an adjoining workpiece. In nearly every

The Mortise-and-Tenon Joint ▪ The mortise-and-tenon is a fundamental woodworking joint that will stand the test of time. The edges of this tenon have been rounded to fit the ends of the routed mortise.

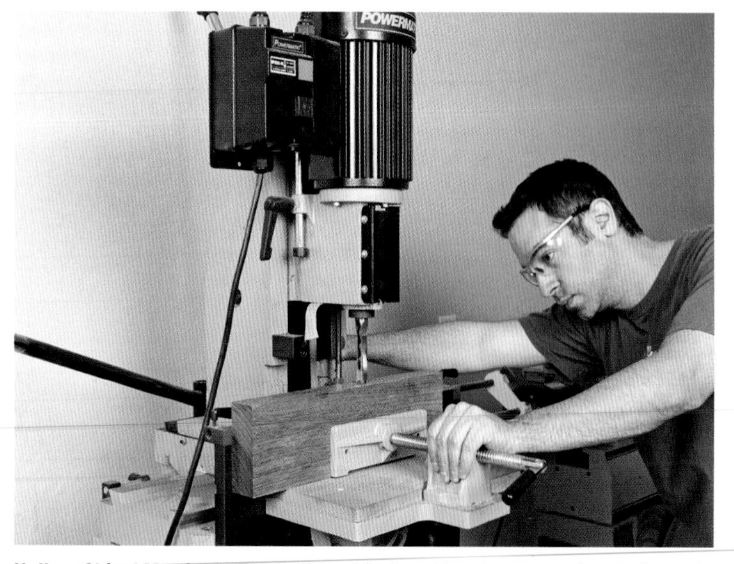

Hollow-Chisel Mortiser ■ The hollow-chisel mortiser is an amazing dedicated mortising tool, but might not be necessary if you learn to use your router.

Spiral Bits are Best ■ When routing mortises, a good quality spiral bit will do the job cleanly and efficiently.

circumstance, I like to cut the mortise first. Why not cut the tenon first? Well, in woodworking you need to pick your battles carefully. Nearly every mortise-and-tenon joint will require some finessing to a perfect fit. Would you rather finesse an exposed tenon on the end of a workpiece, or would you rather try to remove wood from the inside wall of a mortise? There is nothing fun about finessing a mortise. Additionally, mortises are often cut with tools of a fixed dimension. Even if you chop mortises using chisels, the width of the chisel dictates the width of the mortise.

Mortises can be cut with a few different machines, and two of the most common are the router and the hollow-chisel mortiser. The hollow-chisel mortiser operates much like a drill press, only with the addition of a square chisel surrounding the drill bit. The result is something pretty cool: a machine that drills square holes. These machines vary from small benchtop units to expensive heavy duty floor-standing models. I do own one but I don't use it all that often. Whenever I need to cut a mortise, I routinely reach for the router instead. The router requires less maintenance, less setup, makes cleaner mortises and, because it's a hand-guided tool, it can cut a mortise on any size workpiece. The only real advantage I see with the hollow-chisel mortiser is that it creates square-ended mortises. It takes only a few seconds to round the edges of a square tenon to match the mortise, so that isn't much of an advantage. Because most woodworkers already own a router, the hollow-chisel mortiser is by no means a necessity, it's just one more option. I'll focus on the router here.

Two Ways to Make a Mortise

There are two basic ways to rout a mortise, with the machine hand-guided or mounted in a router table. Both methods require a reliable router and a sharp router bit. Not all router bits are created equal and I am a big believer in buying quality. For mortising I use premium quality spiral upcut bits. The spiral shape cuts cleanly and efficiently, and the upcut orientation pulls dust and chips out of the mortise. This reduces heat and as a result, extends the usable life of the bit. Additionally, the tip of the bit has cutting spurs that allow for easy plunging into the wood, creating a clean, flat-bottomed hole. While premium router bits are expensive, they will save you money in the long run. Premium bits don't need to be replaced very often, plus you get clean, quality cuts every time.

Hand-Guided with an Edge Guide

On large workpieces, such as table legs, I prefer to use a hand-guided router. Because routers are top-heavy, it's critical that the router base have full support throughout the cut. You can do this using scrap or by butting up multiple workpieces. Add an edge guide to prevent the router from wandering; most times, the manufacturer's edge guide is adequate and yields decent results. If that's not the case with your router, consider an after-market edge guide with micro-adjustment features, or make your own from wood and threaded rods. The edge guide allows you to push the router along the length of a workpiece while keeping the bit a fixed distance from the edge. As long as you apply pressure consistently, the bit will run true and the result should be a straight mortise of consistent depth.

Edge Guide for Hand-Guided Routing ▪ The edge guide bears against the workpiece to help guide the router when making mortises.

While a fixed-base router can mortise, the plunge router is better because you can adjust plunge depth on the fly. It's usually best to work your way down in small bites. Plunge routers have indexed stops that you can set ahead of time, so after each pass you simply plunge down to the next stop and go. Once you hit the final stop, you're at the desired depth. So if at all possible, get yourself a decent quality 2¼ hp plunge router and you'll never regret it.

I lay out the mortise very carefully, using a sharp pencil. If I have multiple identical mortises to cut, I do the full layout on one piece, and extend the start and stop lines across the other workpieces. With the edge guide set and locked, I only need to know where to start and stop the router. This makes setup and layout fast and easy.

As most woodworkers find out the hard way, routers sometimes have a mind of their own. The problem typically results from a bad decision clashing with unavoidable physics. To be safe, it's best to first plunge completely at the beginning of the mortise and again at the end. If the router is going to catch the wood in a bad way, it will usually happen at the beginning or the end of the cut as you try to hit that pencil line perfectly. So get that step out of the way first while you still can see the pencil lines clearly. After a full-depth plunge at each end of

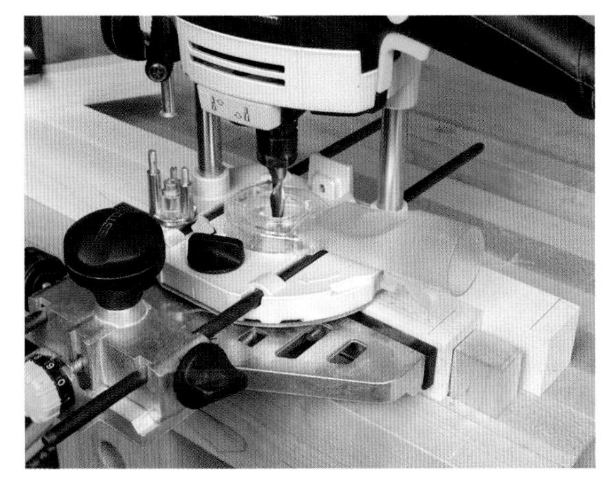

Support is Key ▪ Support the router on a narrow workpiece by butting two of them together.

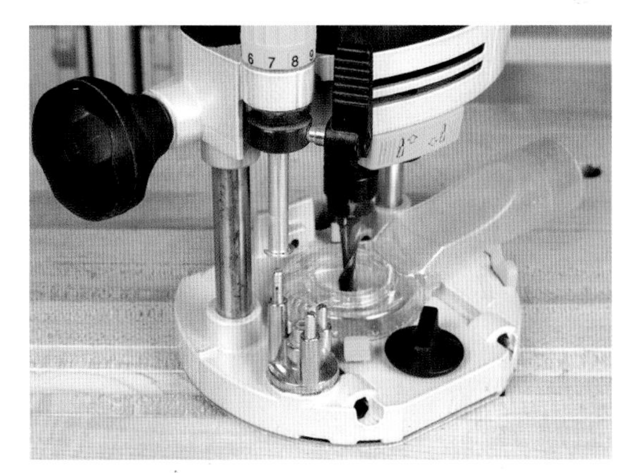

Control the Depth ▪ Plunge routers typically come with depth stops to control the depth of cut when making mortises.

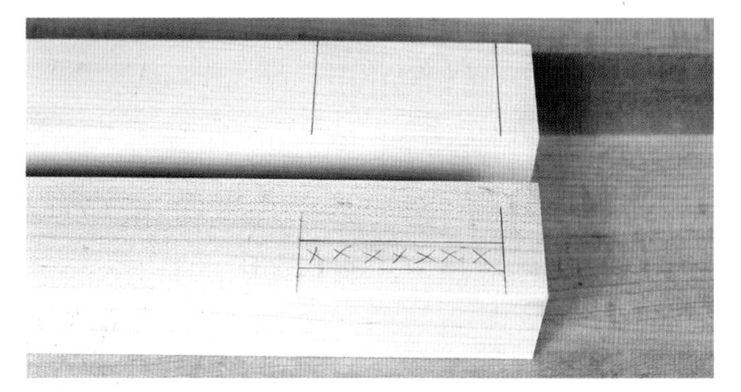

Good Layout is Crucial ■ Careful layout shows everything you need to know to rout the mortise. With one piece marked out, extend the lines across the other workpiece(s).

Start and Stop Points ■ Because it can be tricky to see the start and stop points of a mortise while routing, start by plunging to full depth at the beginning and the end.

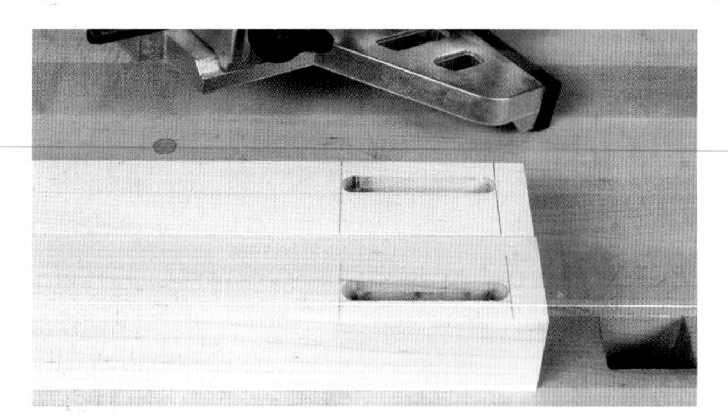

Route the Space Between ■ With the start and stop points established, it's a simple matter to remove the wood that lies between.

Square Off or Not? ■ The round-end mortises can be squared off using a chisel, as an alternative to rounding the tenons.

the mortise, retract the bit and bring it back to the first hole, then begin removing the stock in between, working from left to right with each pass. With a good bit, the mortise walls and bottom should be nice and smooth.

If you have trouble seeing the start and stop lines or you find it difficult to plunge the router bit directly on the pencil line, try using stop-blocks. With the motor off and the bit plunged so that it is almost in contact with the surface, line the bit up with the pencil line at the end of the mortise. Slide the stop over until it contacts the router base and clamp it to the workpiece. Alternatively, you might find it easier to position a stop so that it contacts the edge guide instead. Repeat this operation on the other end of the mortise. With stop-blocks, there's no eye strain and no risk that you'll make the mortise too long or too short. If you have a bunch of mortises to cut and you prefer the stop-block method, you can devise a jig with built-in stops. This would be the best way to ensure accuracy from mortise to mortise.

One side effect of routing mortises is that the ends of the mortise are rounded instead of square. Because tenons are typically square, you might be tempted to use a chisel to square off the ends of the mortise. That will certainly get the job done and you're welcome to do it, but if I wanted to chop a mortise by hand, I would have done it that way from the start. Instead, I round the tenon in best hybrid fashion, using a rasp (page 61).

Router Table for Small Frames

For smaller workpieces such as door frames, I prefer to mortise on the router table. The reasons are safety, control and accuracy. When the mortise is located near the end of a narrow workpiece, it presents the substantial challenge of balancing the router. Rather than cobble together an elaborate jig, take the wood to the tool instead of taking the tool to the wood. At the router table, you can carefully drop the workpiece onto the spinning bit and push it

Mortising on the Router Table

Lay Out One Completely ▪ Only one board needs the complete layout. The rest receive only the start and stop lines.

Line Up by Eye ▪ Using the end marks, line up the bit so that it rests between the layout lines.

Test and Measure ▪ The bit is centered along the workpiece when the material on both sides of the test cut are close to the same thickness.

Mark the Bit's Start and Stop ▪ With a workpiece slid up against each side of the bit, draw lines on the surface of the table marking the start and stop points.

Placing the Stop-Block ▪ Use the layout marks to locate the stop-block for the end of the cut.

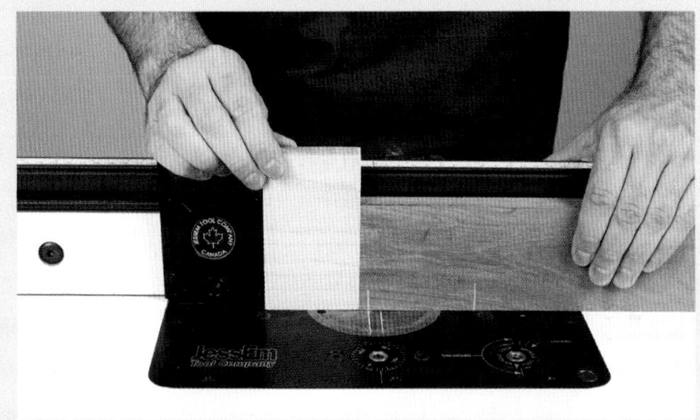

Use a Spacer ▪ Fit a small spacer block against the stop to help control the start of the cut.

Continued on next page

Start the Cut ▪ With the spacer in place, lower the workpiece onto the spinning bit.

Rout to the Stop-Block ▪ Remove the spacer and push the workpiece through the bit until it contacts the stop-block, then lift it off the bit.

Work in Stages ▪ No need to reach full depth in one shot. Take several passes, raising the bit each time.

along the router fence. With guidelines, stop-blocks and accurate setup, this process is both safe and repeatable. So if you have numerous mortises to cut, batch them all at once.

As always, the process begins with careful layout. Mark up one mortise completely, extending the layout lines across the face and end of the workpiece. First, position the fence. With the router bit extended and the workpiece against the fence, align the bit using the layout lines on the end of the wood. Get as close as you can by eye, then run a test piece. You don't need to be scientific about it, but I like to use a set of calipers to measure the remaining material alongside the mortise, to determine if the bit is truly centered on the workpiece. If it isn't, adjust the fence and test again. Once the bit is centered, lock the fence.

Next, place two marks on the router table surface indicating the start and the stop points of the router bit itself. There'll be zero visibility when you drop the workpiece onto the bit, so these lines represent the bit's location. To make the lines, slide a square workpiece along the fence until it contacts the bit, then use a sharp pencil to draw a line against the workpiece. Repeat the process on the other side of the bit.

With layout lines on the faces of the workpieces and start/stop lines on the router table, this method can be very accurate. But to make things even more dummy-proof, add a stop-block. My router fence comes with a fancy stop-block that slides in a T-slot, but you can use a piece of scrap wood and a clamp. Set the stop-block so the end of the mortise aligns with the edge of the router bit. But what about the start of the mortise? Instead of relying on your ability to accurately drop the workpiece onto a spinning bit, you can insert a spacer that will dummy-proof that part of the setup as well. With the workpiece aligned with the start of the mortise, measure the distance between the stop-block and the workpiece and cut a small piece of wood to fit. So now, have the spacer in place as you start each mortise. Once you've lowered the workpiece onto the bit, remove the spacer block and proceed with the cut.

Mortise Oops

As I can personally attest, accidents will happen in the workshop. When making mortises, it's a common mistake to extend a mortise too far. In the case of a stopped mortises, one might accidentally cut the mortise all the way through the end of the workpiece. Fortunately, there's a simple repair.

Whenever possible, I use wood for repairs. Wood fillers and epoxy will get the job done in some situations but there really is nothing more natural looking than a wood repair. This is why I always keep all scraps and off-cuts, no matter how small, until the project is complete.

In this case, the mortise was cut so long it extends all the way through the end of the workpiece. To patch it, I found a piece of scrap stock and planed it to size with the thickness planer. The goal is for a snug fit, but not one that is so tight it pushes the sides of the mortise apart. I then cut the scrap piece to length, coated it with glue and inserted it into the mortise. Clamps hold everything in place until the glue dries.

A few hours later, I planed the patch flush with the block plane, followed by a light sanding using a block.

As you can see, with careful attention to color and grain direction, a patch such as this can be nearly invisible. With the mortise patched up, you can proceed with re-cutting it to the correct dimension.

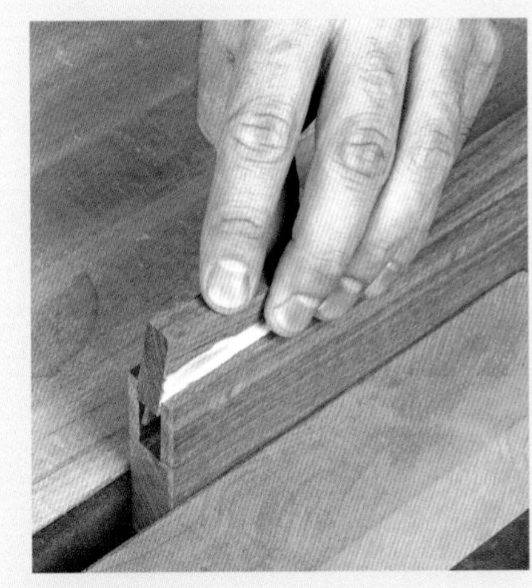

Patch it Up ▪ Find a similar piece of scrap, mill it snug and glue it in.

Clamp Securely ▪ Clamps keep the walls from bulging when the patch goes in.

Plane the End Grain ▪ Once the glue is dry, plane the end grain flush.

Plane the Long Grain ▪ Plane the long grain for a flush fit

Smooth with Sandpaper ▪ Finish off the patch with a light sanding using a block.

Patched and Ready ▪ Once it's patched up, rout the new mortise.

Scribe the Shoulders ▪ A cutting gauge is perfect for scribing the shoulders onto the workpiece.

Lay Out the End Grain ▪ For reference, lay out the tenon on the end grain. On dark woods, use a white pencil.

The last setting is the bit height. I never cut a full-depth mortise in one shot. That would put a lot of stress on the bit and present some safety issues, more so when dropping the workpiece onto a spinning bit. Cut no more than ½" deep at a time. So for a 1"-deep mortise, make the cut in two passes.

This is a very low-tech method for achieving absolutely perfect repeatable mortises.

The Tenon

With the mortise already cut, you now have one less variable to be concerned with because you know the exact recess the tenon needs to fit. As a bonus, you have the actual mortise for test-fitting. The hybrid cut-to-fit methodology removes ambiguity from the process because the answer to the question "Is the tenon the correct size?" is now clearly either "Yes" or "No."

Tenons can be cut with many different tools including the band saw, the router table, the radial-arm saw and, of course, the table saw, my preferred method. Outfitted with a dado stack, the table saw can make quick work of removing stock while producing nice, crisp tenon shoulders. But before firing up the saw, you'll need to do a little layout work at the workbench. Using a knife-style marking gauge, score the tenon's shoulder lines on all four sides of the workpiece. The depth of the mortise dictates the shoulder location. This scribe line will not only help set up the table saw, it will also serve as an insurance policy against accidental tear-out by cleanly severing the cross-grain fibers along the shoulder. Also take time to lay out the tenon dimensions on the end grain of a sacrificial test piece, milled to the exact dimensions of the actual workpiece. It's always a good idea, whenever cutting joints, to use a piece of stock dedicated to setup and testing.

Test Pieces

Whenever I am setting up for joinery, I always make a sacrificial test piece. This is not something you want to think about just before cutting the joints. It's much easier to account for it during the rough-cutting/milling phase. For every project piece, I usually include enough stock to make at least one extra part for testing purposes. This way, the test piece is carried through the entire milling process. By the time you're ready to cut joints, you have perfectly matched test pieces that are the exact same dimensions as the actual workpieces. Because most joints are at the ends of boards, I leave test pieces a bit longer than the actual workpieces. After making the setup and test cuts, I usually still have enough stock to use the test piece as a spare should something go horribly wrong with an actual workpiece.

Test for Perfection ▪ Test pieces are crucial to your success in proper machine setup.

You can't cut the actual workpieces until after you confirm the setup.

Dado Stack

The dado stack can remove a lot of material in a hurry and when making tenons, it saves a huge amount of time. For the average tenon, I set the dado stack for a ¾"-wide cut. The width of the dado isn't critical as long as it doesn't exceed the length of the tenon. If the dado stack is wider than the tenon's length, the blade will interfere with the fence or the stop you'll be using to locate the final shoulder cut.

I prefer to saw tenons using either a miter gauge or a shop-made dado crosscut sled. When using the miter gauge, I use the table saw fence as a stop. When using the crosscut sled, I clamp a stop-block to the sled's fence for a simple and repeatable cut. Both methods are valid and which one you choose really depends on your personal preferences and what fixtures you have in the shop. I prefer the crosscut sled.

Using the test piece, set the blade height at least 1⁄16" below the pencil line on the end grain. Make a quick cut on each face right at the end of the test piece and check how it fits into the mouth of the mortise. Because the blade height was intentionally set low, the fit should be too tight. Raise the blade slightly and repeat the cuts. A second test-fit will be closer to the goal but not quite exactly there. A third adjustment usually hits the mark.

The Right Fit

So what are we aiming for here? In the power tool-exclusive world, we would shoot for a perfect fit right off the saw. But the problem with that is, no matter what we do, there will always be slight variations to contend with. The workpieces might vary in thickness and/or squareness. You might not apply perfectly consistent downward pressure, the workpiece might sit higher or lower during the cut. The dado blade's momentum might lift the workpiece ever so slightly. A stray wood chip could cause the workpiece to register improperly. You also might find that even though the test tenon appears perfect, some subsequent workpieces might end up a little too tight or slightly loose. That's why our goal at the table saw is a fit that by normal standards would be a bit too tight. You should just barely be able to get the corner of the tenon into the mortise, noticing that if you continue to push, you will likely do some damage. That's the sweet spot. When you have a significant number of tenons to cut, the hybrid methodology might be somewhat impractical. In that case, I recommend using the

Miter Gauge and Fence ▪ Tenoning setups don't need to be complicated. A miter gauge and the rip fence can do the job effectively.

Crosscut Sled ▪ On a crosscut sled, a simple stop-block ensures accurate and repeatable tenon cuts.

Raise the Blade ▪ Guided by the end-grain marks, raise the blade, aiming to be just below the pencil line.

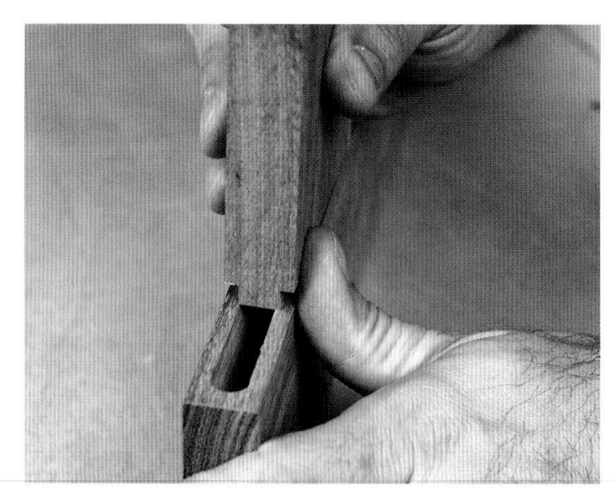

Test the Fit ▪ The initial test-fit should be too tight. Raise the blade slightly and cut again.

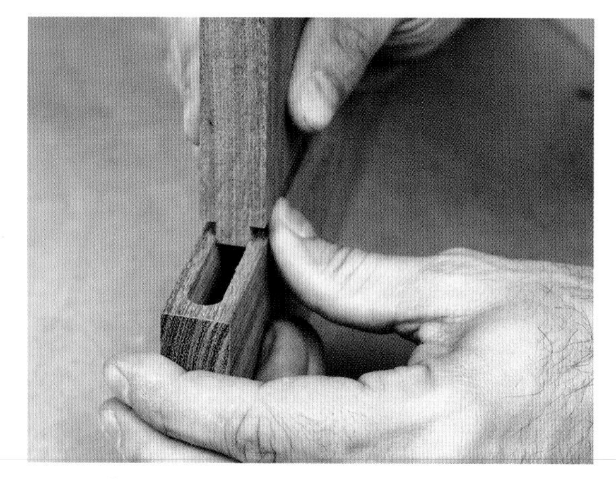

Test Again ▪ The goal is a fit that's just a bit tight, leaving material for some finessing at the workbench.

A Perfect Shoulder ▪ The shoulder cut should just remove the cutting gauge scribe marks. Pictured at left is a cut that's too far away. The example on the right is perfect.

Cut the Cheeks ▪ With the stop-block in place and the blade height set, cut each side of the board to expose the tenon.

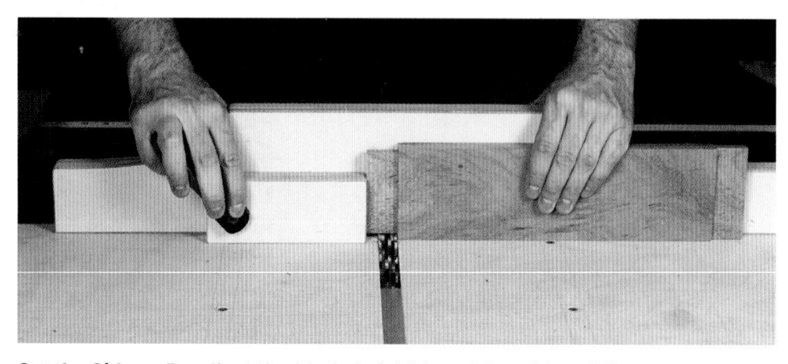

Cut the Sides ▪ Readjust the blade height to cut the sides of the tenon.

table saw exclusively for the sake of expedience. With careful setup and attention to workpiece positioning and pressure, you can achieve results that would be considered a good compromise.

With the blade height locked in, use the scrap piece to locate the stop that marks the tenon shoulder, either the stop-block or the table saw fence, depending on your setup. Much like the blade height, I like to sneak up gradually on this setting. If you used a knife-style marking gauge to establish the shoulder line, you can inspect the shoulder after each test cut to see whether the wood has been completely removed all the way up to the scribe line. If you see a slight ridge at the top of the shoulder, it means there is still a little bit of material to remove. The visible ridge is actually the scribe line itself and our goal is to see no ridge at all after the cut. A few test cuts should be all it takes to get the ideal setting.

With the setup locked in, cut the actual tenons. Start by cutting at the end of the workpiece, making your way closer to the shoulder with each additional pass. For the final shoulder cut, to help establish clean and consistent shoulder lines, rest the workpiece firmly against the stop. Flip the workpiece over and repeat the process from the other side. Make this same series of cuts on all of the workpieces that require the same tenon. Keep in mind that most tenons also require short-shoulder cuts on the edges of the board. Fortunately, all you need to do is adjust the blade height. Because the

Watch the Fuzzies

Nearly all freshly cut boards have fuzzies, the tiny bits of wood that are too small to be called tear-out, but too big to be ignored. Because the fuzzies exist at the end of the board, they can get in the way of perfect registration against stop-blocks. Even if they only impact the work by 1⁄64", that's enough to make your tenon inaccurate and your life difficult during final assembly. The easiest way to combat improper registration is to remove the fuzzy material. I do it by making a few light passes with a shop-made sanding stick. I don't apply a lot of pressure because I'm not looking to create a round-over or chamfer. I'm just knocking down the wayward wood fibers to create a clean corner.

Another precaution is chamfering the edges of stop blocks. This way, if a workpiece should have some fuzzies, they won't contact with the stop block thanks to the small cavity created by the chamfer.

Knock Down the Fuzzies ▪ Knock down fuzzy wood fibers using sandpaper on a stick.

Stop-Block Chamfers ▪ Chamfer stop-blocks, creating a recess for frayed wood fibers.

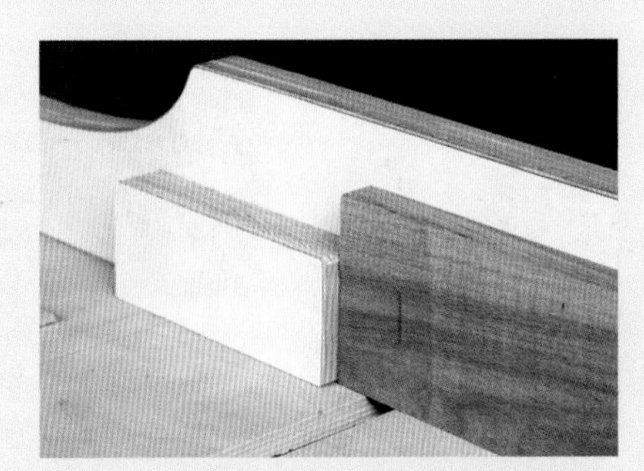

Perfect Registration ▪ With no fuzzies and some clearance, the workpiece registers perfectly.

Using the Fence as a Stop with the Miter Gauge

When making a crosscut at the table saw, the work must be fully supported during the entire cut and the cut-off must not fall between the fence and the blade. A small piece of wood can quickly turn into a bullet if it contacts the fence and a spinning blade at the same time. This is exactly why you'll hear many folks echo a general rule that says, "Never use the miter gauge in conjunction with the table saw fence." This is a statement I wholeheartedly agree with, but only when the cut results in a cut-off. When we make tenons, the waste product is sawdust, not a small block of wood, so there is nothing to get wedged between the blade and the fence. That's why it is perfectly safe to cut a tenon using the dado head, the fence as the stop and the miter gauge to push the workpiece through.

A Dangerous Situation ▪ A cut-off sitting between the fence and the blade is liable to cause a kickback.

Not So Dangerous Situation ▪ Cutting a tenon is safe because there is no loose cut-off between the fence and the blade.

shoulder is the same distance from the end of the board on all four sides, the stop-block will not need to be adjusted for the short-shoulder cuts.

Finessing the Joint

With the heavy lifting complete, it's time to retreat to the workbench to finesse each joint to perfection. You probably have multiple tenons to work on, and it is likely that they all need the exact same treatment. The same number of strokes with the same tools on the same faces could very well yield a set of equally perfect joints. But because of the numerous variables, it's important to test-fit each tenon in its own mortise. Before getting on with the process, I label all the parts so that each tenon is officially partnered with its mating mortise.

Stay Organized ■ When finessing joints, match each tenon to its mortise. I use numbers and letters.

Out-of-Square Shoulders ■ An out-of-square shoulder will result in an unsightly gap in the final assembly.

Square Shoulders ■ With proper machine setup, the tenon shoulders should be nice and square.

First, examine the workpiece to get a feel for what needs to be corrected or finessed. Start by examining the most visually critical part of the joint: the shoulders. When a mortise-and-tenon joint has been assembled, it's the meeting of the shoulders and the adjoining workpiece that everyone will see – and it's all that they will see of the joint. So if the shoulder is out of square or damaged in any way, it will be obvious in the finished piece. Assess the situation with a small square. Are the shoulders square to the edge and the face? Are all four shoulder surfaces in the same plane? Are the short shoulders undercut or sitting proud of the long shoulders? Believe it or not, this last occurrence is quite common when tenoning at the table saw, typically because of out-of-square fences and stops.

To examine the situation in more detail, let's revisit the cutting process. When cutting the long shoulders with the workpiece lying flat, the work is registering off the bottom of the fence or stop-block. When cutting the short shoulders with the workpiece on edge, the work registers off whichever part of the fence or stop-block is sticking out the farthest, whether it be the top, the bottom or, in the case of a square fence, the entire surface. If the fence or stop-block is ever so slightly out of square, tilting ever so slightly toward the blade at the top, this slight deviation will register the workpiece differently in each cutting position. As a result, the short-shoulder cuts will wind up pushing the work away from the blade, producing slightly raised short shoulders. Fortunately, the hybrid woodworker can easily fix these problems.

Out-of-Square Shoulders

If the shoulders are out of square, take some time to investigate which parts of the shoulder are truly out of square and where material

A New Square Reference ▪ Use a square and a sharp pencil to draw a square reference line all the way around the board.

Plane to the Line ▪ A shoulder plane is the perfect tool to fix out-of-square shoulders.

will need to be removed. Use a sharp pencil and a square to draw a continuous line all around the workpiece, as close to the shoulder as possible. This line will serve as a reference guide for planing the shoulder. One out-of-square shoulder is likely to be out of square on the opposite shoulder as well, and after a few minutes of planing, it's easy to end up with two perfectly square shoulders that are no longer in the same plane. The pencil line is a reality check to prevent that from happening.

With the workpiece flat on the workbench, use the shoulder plane to pare away the excess shoulder material on both sides. The cheek of the tenon does a great job of supporting the plane, keeping its body nice and square to the shoulder. Most shoulders can be fixed with a few strokes per side. Anything more than that and you're probably better off to re-calibrate the table saw setup and recut all the workpieces.

Proud Short Shoulders

With the long shoulders square and in the same plane, the short shoulders may be a bit proud. Regardless of whether this is due to the table saw setup or the finessing of the long shoulders, the repair process is the same. Typically the short shoulders are so short that the rabbet plane isn't the best tool for the job. Instead, pare with a sharp chisel. Thankfully there are two perfect reference surfaces on either side of the proud short shoulder, so paring the wood away is a breeze. Start with the tip of the chisel

Short Shoulders ▪ Trim the short shoulders with a sharp chisel, using the long shoulder as a reference surface.

The Big Picture

When trimming tenon shoulders, don't lose sight of the big picture. Most tenoned workpieces have a tenon on each end. As you trim the tenon shoulders back, you are effectively shortening the between-shoulders distance. So if the shoulder-to-shoulder distance on one workpiece doesn't match the shoulder-to-shoulder distance on a second identical workpiece, you could end up with joints that don't close completely and an out-of-square assembly. So when trimming shoulders, you can avoid subsequent problems if you compare your workpieces to one another. We finesse tenon shoulders to solve a problem. Ideally, we would prevent this problem in the first place with proper setup at the table saw.

Dado Grooves ▪ A typical dado stack leaves numerous grooves in the surface that can serve as a visual aid.

lying flat on the long shoulder, then slowly and deliberately rotate the blade into the excess wood. With this method, the chisel works much like a plane as it peels away any excess material that sits proud of the long-shoulder surface. When each short shoulder is flush with the two long shoulders, the work is done.

Undercut Short Shoulders

If the short shoulders happen to be undercut, the repair is similar to fixing out-of-square long shoulders, only easier. Use the shoulder plane to take as many passes as needed to bring the long shoulders flush with the short shoulders. When everything is nice and flush, the shoulders should all be square and all in the same plane. Check your results with a square and if you think you need to plane some more, don't be afraid to draw a square reference line all around the workpiece as a little insurance policy.

Tenon Cheeks

With the shoulders of the tenon squared away, turn your attention to the tenon cheeks. In all likelihood, the tenon will be rough with slight grooves created by the dado blades. To cut cleanly, dado blades have beveled teeth that score the work, dramatically reducing tear-out. This is a great feature with an unintended secondary use. Because the grooves are at a fixed depth, you can use them as depth gauges. Most tenons are meant to be centered on the thickness of the workpiece, so if you remove more wood from one side of the tenon than the other, the tenon will no longer be centered. Because you don't have much wood to remove at this stage,

Check the Stop Block ▪ A stop block should be square to the work surface.

A Proud Short Shoulder ▪ If the stop block is not square, the result could be proud short shoulders.

An Ounce of Prevention

If you notice that your short-shoulder cuts are always raised or recessed when compared to the long shoulders, there's a good chance the stop block is to blame. If the end of the stop block is not perfectly perpendicular to the surface of the saw table or the cross-cut sled, you will be plagued with mismatched shoulder cuts. The reason is, the workpiece registers at different parts of the stop block for each cut. With the workpiece laid flat, it registers on the bottom of the stop block. With the workpiece on edge, it registers on whichever part of the stop block sticks out furthest. If the stop block isn't square, these two cuts will always be slightly different. So before cutting tenons, use a square to make sure the stop block is square and perpendicular to the surface. And of course the workpiece itself should also be square, otherwise you can find yourself in the same predicament.

these shallow grooves can serve as the perfect visual reference for finessing the fit.

Two tools are ideally suited for finessing tenon cheeks: the rabbeting block plane (I prefer this over a shoulder plane), and the router plane. I prefer the wider body of the rabbeting block plane over the shoulder plane, but either tool will do a good job here. With the workpiece flat on the bench, take a light pass with the plane pushed hard up against the shoulder. To help prevent tear-out, slightly chamfer the long-grain side of the tenon with a chisel. After taking a pass on each of the tenon's cheeks, do a test-fit. Because the tenon is square and the mortise has round ends, the tenon will never fit in completely. But it's easy to judge the thickness of the tenon by pushing its corner into the center of the mortise. Sometimes the tenon will fit on the first try. If that's the case, pat yourself on the back and go celebrate with a cup of coffee. If not, grab the rabbeting block plane and take another pass or two, the same number of strokes on each side with an eye on the depth-gauging grooves. It's OK if you don't completely remove the grooves by the time the tenon fits, no one will ever see it. But the grooves should be less noticeable as the tenon slides ever so sweetly into its mortise.

The router plane is the other tool I like to use to finesse tenon cheeks. Unlike many other planes, the router plane has a depth stop and can only cut to its pre-set fixed depth, making it the dummy-proof choice. With the workpiece flat on the workbench, set the widest square blade you have for a very light cut and lock its depth below the plane sole. Keep downward pressure on the workpiece side of the plane body while you push the whole assembly forward and back, removing any material that's in the way. When the blade stops cutting, the work is done. Flip the piece over to repeat the process. With the router plane, there is no need to count strokes, like there is with a rabbeting block plane because the tenon will always end up centered on the workpiece. As a bonus, once the router plane has been set it can be used on every workpiece. As long as the workpieces are all the same thickness, you

Finesse the Cheeks ▪ A rabbeting block plane is a great tool for cleaning up a tenon cheek.

Test the Fit ▪ The tenon has square ends, but you can still test the fit by pushing it in at an angle.

The Router Plane Option ▪ Router planes, with their easy depth adjustment, excel at cleaning up tenon cheeks.

Balancing the Router Plane

If you are new to the router plane, you might find it tricky to balance the tool at the end of a workpiece while focusing pressure on the shoulder side of the plane. With more than half of the plane's body hovering over the tenon cheek, it doesn't take much for the plane to tilt slightly and remove more material than you intended. Fortunately, there are two tricks you can employ to stack the cards in your favor.

Extend the Base ■ Attaching an oversized sub-base to a router plane makes it easier to balance over the end of a board.

The Sub-base Extension

To help counterbalance the plane, attach a thin piece of sheet stock to the base of the router plane. The sub-base should extend beyond the sides of the plane so that the effective reference surface is increased by an inch or two. With just a small amount of additional hand pressure on shoulder side of the plane, it should be easy to keep the plane balanced, level, and perfectly positioned over the tenon. If you want to get fancy, you could attach a small block of wood to the sub-base as a handle for additional control.

The Two-Tenon Trick ■ By butting two tenons end to end, you can trim both at once with full support on both sides of the router plane.

The Two Tenon Trick

Instead of working on one tenon at a time, try butting two identical tenons together. With two workpieces flat on the workbench tenon-to-tenon and immobilized with clamps, the router plane will have enough support on both the left and right side, making it much easier to balance. Furthermore, you get a two-for-one deal.

Rounding with a Rasp ■ A rasp will make quick work of rounding the edge of a tenon to match a round-ended mortise.

can quickly and accurately size each tenon to perfection.

Rasping the Tenon Round

Because router bits are cylindrical in shape, routed mortises have rounded ends. In contrast, right off the table saw, tenons have square ends. Instead of squaring up the mortise, which takes more time and effort than I want to invest, I round over the tenon to match. I used to chisel the tenon in an attempt to perfectly match the mortises, but I spent way too much time on such a simple task. Furthermore, the grain of the wood would often make it difficult to pare from the tip of the tenon toward the shoulder, tear-out was common and the risk of splitting was great. Thankfully, a reader of my website pointed out that I should instead be using a rasp for this job. While not exactly a tool of great precision, a small rasp can round over the tenon in mere seconds, plus rasps aren't much affected by grain direction. I use a chisel to sever the fibers where the tenon meets the shoulders, but the rasp does the rest. With a little practice, just a few strokes is all it takes to round the edges of the tenon, allowing it to fit the mortise like a dream. I'll never go back to the chisel again.

The loose mortise-and-tenon is another option to consider, which by its nature addresses the rounded mortise issue. Instead of the tenon being an integral part of one workpiece, you rout matching mortises in both adjoining pieces, then mill the tenon stock separately. Using this method, you can round over the tenon stock with a router bit, then crosscut it to length and insert the loose tenon into both mortises for a perfect fit.

Trimming the Tenon to Length

Another trimming operation that might be required usually doesn't come to light until after you've completed everything else, and that's shortening the length of the tenon. After inserting the tenon as far as it will go, a consistent gap between the tenon shoulder and the mortised workpiece indicates the tenon is too long and bottoming out. If the gap is more than 1⁄16", I use the miter saw to trim the tenon to length. If the gap is less than 1⁄16", I prefer the block plane. Clamp the workpiece vertically and low to the workbench. If there is too much distance between the clamp and the tip of the tenon, the workpiece will vibrate, chatter and perhaps squeal when planed. It makes for unusual music but doesn't do a whole lot for the wood. So mount the piece low and clamp as close to the tenon as possible. Set the block plane for a fine cut with its adjustable mouth only allowing a fine shaving to pass through. As an additional precaution, use a chisel or rasp to lightly chamfer the far end of the tenon where tear-out is always a risk. After just a few passes, it's time for a test-fit. If the gap persists, it's back to the workbench for a few more passes. The work is not complete until the shoulders touch evenly on all sides of the joint with no space at all.

A Snug Fit ▪ With both edges rounded, the tenon should slide nicely into the mortise.

Too Long ▪ If the tenon is too long, plane it to length so it doesn't bottom out in the mortise.

Soften it Up ▪ If you're having trouble planing end grain, try softening up the fibers using denatured alcohol.

Assembled Joints ■ This fully assembled frame features four mortise-and-tenon joints.

Assembly

The assembly phase of a project is where the rubber meets the road. Errors in milling or joining will rear their ugly heads at this stage, so it's important to do a dry assembly. Let's use a door frame with four mortise-and-tenon joints as an example. After testing the fit of each joint individually, making sure the parts remain square and flat after assembly, put the entire frame together using light clamping pressure. Is the frame still square? Check each corner inside and out, using a reliable square. You can also check for square by measuring from corner to corner. If the frame is out of square, the two corner-to-corner measurements will not be the same. If that's the case, you'll need to do some investigation to figure out why. Sometimes the frame just needs a little persuasion in one direction or another to square it up, but other times, there is something amiss.

Determining exactly what went wrong and how to fix it is beyond the scope of this book, but I have good news for you: a good mortise-and tenon-joint is mostly self-squaring. Assuming the workpieces were milled consistently and the joints were cut accurately, the construction will square up upon assembly. Because the shoulder is perpendicular to the tenon itself, if the shoulder is bottomed out and gap-free, the tenon workpiece should be perpendicular to the mortise workpiece. So when a dry-assembled frame is out of square, the first thing to do is look at the shoulders of each joint to see if a small gap might be present. Usually a small amount of pressure applied corner to corner will quickly square everything up.

The Perfect Fit

Throughout this discussion I've referred to a "perfect fit." But what exactly is a perfect fit? In my experience, the perfect fit is when the tenon slides into the mortise with some effort, but not so much effort that you need the help of a mallet. When you turn the assembly upside down, the piece should not fall out due to gravity alone, but it could be removed by hand. This is a slip fit. Anything looser leaves too wide of a gap between the tenon and the mortise walls. Anything tighter will be difficult to assemble, made worse after the wet glue swells the wood. Furthermore, the glue gets scraped off the tenon during insertion into the mortise, and the result could be a glue-starved joint. A slip fit remains easy to assemble even after adding glue, and leaves an ideal space for glue between the mortise wall and the tenon.

A Slip Fit ■ A slip fit means you don't need a hammer to put it together and you need more than the force of gravity to pull it apart.

Glue-up

Now disassemble the frame for the final glue-up. If the dry assembly went well, there shouldn't be much in the way of surprises once you add glue. I put a light coating of glue on the tenon cheeks and sides, but I avoid putting glue on the shoulders. There isn't much glue strength between the shoulder and the area surrounding the mortise, because this is an end-grain to long-grain bond, so leaving the glue off avoids excessive glue squeeze-out. I also add a thin coat of glue to the mortise walls. With glue on both surfaces, the joint should slide together easily, although you might notice things are tighter due to the moisture in the glue swelling the wood fibers. Use two parallel clamps for a simple frame, one across each joint. If there is any glue squeeze-out, let it dry for an hour and skin over. Then the droplets of glue can be scraped away without spreading the glue all over the surface.

Level the Joints ▪ Each frame joint may need a small amount of scraping or planing to level the surface.

Flattening

Once the glue has completely cured, inspect each joint closely. The seam between the workpieces will almost never feel perfectly flush after glue-up. A light scraping with a card scraper followed by a quick #220-grit sanding is all it takes to make the joint feel perfectly smooth. In some cases, however, you might notice that the surfaces are slightly offset from one another. You can quickly remedy this using a cabinet scraper, a block plane or a smoothing plane. Because the door frame is at finished dimension, it's important to read the grain carefully before planing because tear-out at this stage would be disastrous. That's why I favor scrapers for this task. Slight variances are nothing to worry about but an offset greater than 1/16" could very well interfere with the fit of the door. At worst, the entire surface of the frame can be worked with a plane so the joints are smooth and the door remains flat, even if it means the door is now slightly thinner than originally intended.

No Over-Cutting

One great advantage of the hybrid methodology becomes clear after making a mortise-and-tenon joint: the degree of control. There is hardly any risk of over-cutting, as long as you're paying attention. When using power tools, a well intentioned adjustment turns into a major boo-boo when too much stock gets removed. When it comes to hand tools, one pass too many hardly ever spells disaster. When set for a light cut, a plane removes only a few thousandths of an inch at a time, so even when you go over the target, you haven't gone very far. In my opinion, a few thousandths beyond the target is still a bull's-eye. But keep in mind that we are always trying to get as close as possible using power tools first. Most hand-tool operations should take no more than three or four strokes to reach the desired state. More than that is a clear sign that you should have been more aggressive with power tools in the first place. The increased accuracy conferred by hand-tool finessing is well worth the extra time it takes. But I certainly don't want to add any more time than is absolutely necessary. The goal with power tools is always, like the old game show says, to get as close as you can without going over. I don't want to make a tenon at the workbench, I only want to finesse the tenon I've already made.

Fixing Tear-Out

When planing end grain, tear-out is always a concern. This applies to trimming shoulders as well as trimming the end of a tenon. While a bit of tear-out on a tenon isn't a major issue (because it will be hidden inside the mortise), tear-out at the tenon shoulder will create an unsightly flaw. Chamfering the end grain with a chisel stacks the cards in your favor, but tear-out can still occur, so it's important that you know how to fix it.

Because we are using hand tools, the torn-out pieces probably are still attached to the workpiece or they are sitting on the workbench or floor. Recovering the chunk of material is essential for a perfect repair because you can reattach it using thick or medium cyanoacrylate glue. I use a small piece of scrap wood to apply pressure while the glue dries. I also use a quick-set activator to speed up the curing process. Once the glue has dried, I use a scraper to remove the squeeze-out and smooth the surface.

A Common Mishap ▪ Tear-out at the shoulder is a preventable but common occurrence when finessing a tenon.

Reattach It ▪ Using cyanoacrylate (CA) glue, reattach the torn-off piece of wood, or find a comparable substitute.

Don't Glue Your Fingers ▪ CA glue dries quickly so I use a small piece of scrap to hold the repair in place.

Flush it Up ▪ A card scraper and a little sanding makes the repair flush and invisible.

Half-Laps ■ Half-laps come in a few different varieties, including the corner half-lap and the cross half-lap shown here.

Half-Lap Joints

While a mortise-and-tenon is the joint of choice for most structural applications, there are many situations where a simple half-lap joint would do the job with less effort and in less time. Half-laps aren't as strong as the mortise-and-tenon and have little resistance to racking forces, but they are certainly strong enough to use in frame-and-panel doors and other small frame projects such as mirrors and picture frames. Just be aware that the wider the workpiece, the greater risk of racking or twisting.

There are a number of variations of the half-lap joint and we'll cover two of the most common here: corner half-laps, and cross half-laps with two variations. As the name suggests, half-lap joints usually involve removing half of each piece's thickness, allowing the two to overlap and nest together with their faces perfectly flush.

The simplicity of the joint is deceiving and many folks find out the hard way that if the pieces aren't cut just right and clamped together properly, the result will be uneven faces and/or unsightly gaps. The hybrid method not only helps cut the pieces accurately, but also allows cutting the pieces in a way that facilitates easy assembly and effective clamping. More on that later.

In the mortise-and-tenon section we discussed several factors that could lead to natural variances in results, and because the procedure for making a half-lap is very similar, those variances apply here as well. By using hand tools to finesse the joints, we can overcome those variances and achieve perfectly predictable results every time.

Corner Half-Lap

I cut corner half-lap joints at the table saw using either the miter gauge or a crosscut sled. The process is very similar to making tenons. Start at the workbench by defining the shoulder lines with a cutting gauge. Instead of setting the cutting gauge to a particular numeric measurement, I use the adjoining workpiece as a guide because its width dictates the location of the shoulder. But I don't actually set the cutting gauge to the exact width of the board. Instead, I set it about 1⁄32" under. By undersizing, the edges of the lap will protrude slightly from the end grain of the adjoining piece and that will be a tremendous help when clamping the joint later. Only one face of each board needs to be scribed. Be sure to also scribe a test piece for the initial table saw setup.

Just Shy on Purpose ▪ Set the cutting gauge about 1⁄32" shy of the width of the adjoining workpiece. This will help when clamping.

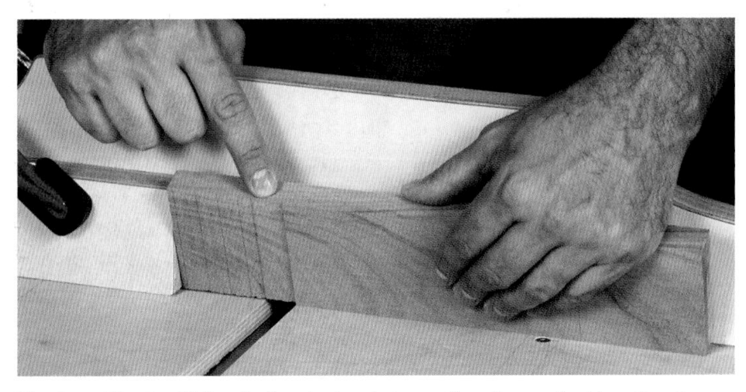

The Stop-Block ▪ Make shallow test cuts to confirm the perfect location for the stop-block.

Cut the Half-Lap ▪ Once the stop-block is in place and the blade height set, sawing the half-lap is simple.

At the table saw, I use a dado stack and either a crosscut sled and stop-block, or the miter gauge and rip fence. The gauge line on the test piece provides the information needed to locate the stop. Make small shallow cuts, moving the stop a little bit each time until the blade cuts just into the shoulder line, but not over.

For half-laps with adjoining pieces of the same thickness (which is usually the case), set the blade height to just under half the thickness of the workpiece minus about 1⁄32". If you have bionic eyeballs and can set the blade 1⁄64" under half, go for it – the more material you leave, the more you'll have to remove later at the workbench. The sliver trick (page 138) is a quick and easy way to set the blade height.

With the setup complete, you're ready to cut the project workpieces. Start removing material at the end of the board and work your way back to the shoulder. Make the final pass with the workpiece firmly pressed against the stop. Cut all the pieces using the same setup, assuming all the parts are the same width and thickness. If one of the adjoining pieces is a different width, you'll have to adjust the stop but the blade height stays the same.

Clean It Up

Do the finessing at the workbench using your choice of a rabbeting block plane, a shoulder plane or my personal favorite, the router plane. The process is the same as finessing a tenon. Check the shoulder for square and use the shoulder plane to make any adjustments necessary. Fortunately, there is only one shoulder per workpiece so the risk of trouble is slim. Next, inspect the cheek. The little grooves left by the dado blade are excellent depth gauges if you're cleaning up with a rabbeting block plane or shoulder plane. If you're using the router plane, set the largest blade for a light cut and clamp the two workpieces end-to-end on the workbench. Let the router plane span the gap, cleaning up both half-laps at the same time. Remember to test the fit after every set of passes and adjust the blade accordingly. When the two pieces nest together with perfectly flush faces, you're done.

Because we intentionally undercut the shoulders, the edge of each workpiece will protrude slightly from the end grain of its mate, which helps make clamping the joint easy and effective. To join the two pieces, you not only need downward pressure to sandwich the two laps together, you also need lateral pressure from the outside edges to close up the shoulder. If the pieces were dead-on flush on the outside corner, the clamp wouldn't have any way to apply pressure and close the joint. Undersizing the shoulder cut leaves a little bit of extra material protruding at the joint, allowing the clamp to push the shoulder closed.

After the glue dries, you can clean up the joint and make it flush. Secure the frame pieces in the workbench vise and use a block plane, a smoothing plane or a scraper to flush each edge to the end grain of the half-lap. If the wood is cooperative, a few passes with a block plane or a smoothing plane will do the trick. For best results, work away from the end grain instead of toward it. If the grain direction of the wood requires you to work toward the end grain, lightly chamfer the vulnerable edge of the lap and work carefully toward it.

When working on something such as a door frame, cut the rails and stiles to the exact desired length as dictated by the door opening, accounting for hinge details and a smidge of general clearance. Because you've cut the parts to exact length, you know that when long grain on all four sides has been trimmed flush with the end grain of the laps, the door will be at the desired dimension.

Fine-Tune the Fit ▪ Use the router plane to fine-tune the fit of the half-lap joint.

Assembly • To assemble a half-lap frame, use four parallel bar clamps with an F-style clamp sandwiching each corner.

A Method to the Madness • Setting the marking gauge for $1/32$" undersize results in extra wood that helps the clamp close the half-lap shoulders.

Clean Up the Edge • After the glue dries, clean up the edges and make them flush using any block plane, bench plane or scraper plane.

Cross Half-Lap

There's another common type of half-lap joint known as the cross half-lap. The joint can be in the middle of one board (shaped like an uppercase T) or in the middle of both boards (shaped like a lower-case t).

Regardless of which version you're making, the cross half-lap is potentially more complicated than the corner half-lap because it's located somewhere in the middle of one or both adjoining workpieces. If the joint is in the middle of the board, you'll have two shoulders to be concerned about instead of just one. The distance between the shoulders must perfectly match the width of the adjoining workpiece, otherwise there will be an obvious gap. Thankfully, there's a hybrid woodworking solution that makes this process easy and accurate. Instead of trying to match the shoulders to the width of the workpiece, which usually results in the fit being too tight or too loose, we'll simply get as close as we can (without going over) and then plane the sides of the adjoining workpiece to fit between the shoulders. To make this work, make sure the shoulder-to-shoulder distance is just shy of the width of the adjoining piece.

t-Shaped Cross Half-Lap

In most cases, a t-shaped cross half-lap will be located at the center of both adjoining workpieces, making setup easy. Use the workpieces themselves to do the layout. With centerlines marked along the width of one workpiece and along the length of the other, line up the pieces and trace the outside edges of the top board onto the adjoining piece.

At the table saw, arrange two stop-blocks on the crosscut sled, one on each side of the workpiece. If you prefer using the miter gauge, clamp a single stop-block to its left side and use the rip fence as a stop on the right. Set the stops so the dado blade cuts just inside both layout lines, making a notch that's about 1⁄32"

Sliver Trick

Here's a cool way to set the blade height without ever using a measuring device. Raise the blade so it is just under half the thickness of the test piece. After a quick pass on each face, you'll be left with a thin strip of wood in the center. Break this piece off, raise the blade slightly, and repeat the cuts on each face. Keep raising the blade a little bit at a time and making test cuts until you're left with a very thin sliver of wood. Because the sliver is so thin, it tends to vibrate into the blade causing it to break off in parts. When you see a few fractured bits and pieces left over after the cut, you know the blade is set properly. The thickness of the sliver represents the amount of material you'll have to remove later during the finessing process. Aim for 1⁄32" at most.

Just Short of Half ■ With the blade set just under half the thickness of the workpiece, make a pass on each side.

Leave a Sliver ■ When the blade is set just right, a cut on each side leaves a thin and tattered sliver of wood.

Centerline Setup

Whenever possible, use actual workpieces to make layout marks instead of tape measures and rules. In the case of a centered cross half-lap, you can use centerlines to orient the pieces, then run a pencil along the outside edges to locate the shoulder cuts. So how do we find the center? Fortunately, all you need to do is approximate. For instance, to find the center along the width of one of these pieces, which we know to be about 3" wide, we can set an adjustable square to roughly $1\frac{1}{2}$" and strike a line from each side. No matter what the exact width of the workpiece is, the center is revealed as the space between the two resulting lines.

This technique can also be used to find center along the length of a board, though you'll likely need a ruler or tape measure to measure in from each end. With the center points of the two pieces marked, the boards can be aligned with the help of a large square. Using a sharp pencil, trace along the outside edges of the workpiece to establish two reference lines that locate the shoulder cuts.

Center Along the Width • Set an adjustable square for the approximate centerpoint. Mark a line from each side; the true center lies in between.

Center Along the Length • Use a rule to mark the center along the length. The same distance measured from both ends brackets true center.

Line Up the Centerlines • Align the width and length centerlines and use the workpiece itself to lay out the cross half-lap shoulders.

Ready to Cut • With the shoulder lines in place, these pieces are ready for cutting.

too narrow. The blade height can be set using the same method described previously for the corner half-lap, using a test piece and the sliver trick (at left).

With the settings locked in, remove the wood inside the lines. If the adjoining piece is the same length, simply cut the half-lap using the same setup. If one workpiece is longer than the other, the stop-blocks will need to be reset for the half-lap on the second workpiece.

After cutting the notches on both pieces, head to the workbench. The pieces should not go together because we intentionally undersized both notches. Using a block plane, cabinet scraper or smoothing plane, begin removing small amounts of material from the edges of one of the workpieces and test the fit as you go.

Two-Stop Setup • With two end-stops to anchor the workpiece on the crosscut sled, the saw blade removes the material between the shoulders.

Trim to Fit • Use a block plane to work down the edges of one piece to precisely fit the other piece.

Test the Fit • The goal is for each piece to slide in between the shoulders of the other with just a little effort.

Finesse Work • The router plane cleans up the cheek of the half-laps. It shouldn't take much to achieve a perfect fit.

A Perfect t-Shaped Cross Half-Lap • With just a little extra attention to detail, the cross half-lap fits perfectly.

Because both notches are too tight, you'll have to flip the piece upside down to test the fit in the adjoining notch.

A word of caution about test-fits: it is very easy to dent the crisp corners of a half-lap joint by trying to force the pieces together. Be careful and only use light pressure. Once the piece slips in with a nice snug fit, repeat the process on the edges of the other workpiece. Because the first piece has already been planed to width, the second piece can be tested by assembling the joint with the pieces in their proper orientation.

Next, the cheeks of the half-lap need to be finessed for a nice flush fit. Because there is support material on both sides of the half-lap, the router plane is the perfect tool for this task. A few passes on each workpiece should be all it takes to bring the pieces together perfectly.

T-Shaped Cross Half-Lap

The T-shaped cross half-lap is essentially a combination of the corner half-lap, with one joint on the end of a board, and the t-shaped half-lap, with the other joint in the middle of a board. The two pieces join to look like a capital T. The method for making both pieces has been covered in the previous sections.

After making both cuts at the table saw, head to the workbench for finessing. Much like the t-shaped cross half-lap, you'll need to remove material from one board to make it fit between the shoulders of the other, only in this situation you'll only need to plane the piece with the end half-lap. Begin removing small amounts of material from both edges, testing the fit until the piece slides in without much effort. To clean up the cheeks, the router plane is my weapon of choice.

Assembly

Assembly of both cross half-lap types is straightforward. The t-shaped joint requires nothing more than a single clamp to squeeze the joint together in the middle. The T-shaped joint requires two clamps: one to sandwich the pieces together and one to close up the shoulder gap. Just like the corner half-lap, the outer edge of the center-cut half-lap workpiece should have some extra material to aid in clamping.

Assemble T-Shaped Cross Half-Lap • Another beautiful half-lap joint, finessed for a perfect fit.

Clean Up

The t-shaped joint should only require a small amount of smoothing on the faces where the two pieces meet. A sanding or a light scraping should be all that is required. If for some reason there is a larger discrepancy, a block plane, smoothing plane or a more aggressive scraper can be used. But with careful attention to setup and after a successful dry fit, any surface variances should be minimal.

The T-shaped joint will require a small amount of planing to remove the overhanging material on the edge of one of the pieces, bringing the edge flush to the end grain of the half-lap joint. Because the end grain is captured between long grain, there are no major concerns about blowing out the end grain.

What I find really compelling about this half-lap technique is that instead of finessing the inside shoulders of the half-lap itself or trying to get them perfect right off the saw, we're shaving the adjoining workpiece slightly narrower. This is a much easier and more predictable endeavor, although it might not be an option that is immediately obvious. Not only will the piece fit perfectly every time, the planing process removes any residual machine milling marks from the edges of the workpiece, minimizing subsequent work prior to finishing. This is a shining example of the hybrid system as it not only makes the process easier, it produces better results while saving time and avoiding frustration.

Auxiliary Base Makes it Easier • Use an auxiliary base and a piece of scrap to balance the router plane on the end of a workpiece.

Plug Options ■ Choices for plugs: tapered face-grain plug before and after installation, left, and a dowel plug before and after installation.

Dowels and Screw Plugs

Although most woodworkers strive to avoid using mechanical fasteners, there's no denying they have their place. Fortunately, there are ways to disguise screw heads so well that the onlooker will have no idea they are there. By countersinking the screw, you can glue in small pieces of dowel stock or, better yet, tapered face-grain plugs, to cover the screw head. Once the glue dries, you do face the challenge of flushing the plugs to the surface without damaging the surrounding material. Using screws is not the only situation where this challenge arises; it also occurs with pegged and drawbored mortise-and-tenon joints.

Be Careful in Plywood ■ Plywood face veneers are thin, so when sanding a plug be careful not to burn through the veneer.

Most modern woodworkers reach for a powered sander to flush these plugs. Unfortunately, if you do that the result likely will be a series of hills and valleys. Sanding pads have some cushion to them, so it's nearly impossible to sand a small point without the sandpaper contacting the surrounding material. You can get away with this in solid wood where a little surface irregularity might go undetected, but in plywood it's game over. You'll quickly burn through the thin surface veneer and expose the wood layers beneath. To avoid this, you need to trim those plugs flush to the surface without even thinking about touching them with a sander. Hybrid woodworkers find that a flush-trim saw is perfectly suited to the task.

Flush Trimming ■ After gluing a plug in place, use a flush-trim saw to cut away the excess material.

Trimming the Plug Flush

The flush-trim saw has a thin blade with no set, so its teeth won't mangle the surface as you push it back and forth. To flush a plug, simply lay the blade flat on the workpiece and raise the handle up. The blade will bend and the resulting downward pressure will keep the blade nice and flat. You also can use your free hand to apply downward pressure directly on the blade, just to make sure you get the closest cut possible.

Pushing the blade forward and back, you'll soon see the plug pop off, leaving a nearly perfect surface. The flush-trim saw has small teeth so you can't rush the process. If you were to rush, you would end up with extra material sitting proud of the surface and you'd probably want to make a second pass. With the plugs trimmed, the surface needs nothing more than a light sanding. Although it is safe to use a powered sander at this stage, you'll find the best results come from a sanding block wrapped in sandpaper. This is especially true if you're trimming end-grain dowels, because the end grain will always sand slower than the face grain. The hard surface of the sanding block helps you sand the plug thoroughly without affecting the surrounding material.

Sand it Smooth ■ Sandpaper on a wood block quickly brings the plug flush with the surface.

Edge-Banded Plywood ■ Edge-banding, when executed with care, is a great way to dress up the unsightly edges of plywood.

Edge-Banding Plywood

To some people plywood is a four-letter word. This is unfortunate because when used with discretion, plywood can be a valuable asset. It's an essential building material that satisfies the specific need for flat and stable stock, and I'm thankful to have it. Regardless of how much we use the material in our work, most folks will agree that plywood edges are ugly. That's why it's imperative to hide the edges, either with joints such as dados or rabbets, or by covering them with solid wood or veneer tape. Even if you like using plywood, you probably want to make the project appear as if it was made from solid wood.

While veneer tape is a quick and easy solution, especially when using the pre-glued iron-on type, I prefer shop-made solid-wood trim. For most applications, I use a ¼" strip of solid wood that matches the species on the plywood face. With ¼" of material to work with, I can add a small edge profile, such as a chamfer or roundover, that wouldn't be possible with veneer tape. This detail further supports the illusion that the piece was constructed from solid wood.

Because plywood has such a fragile top layer of veneer, typically only 1⁄32" thick, you have to be very careful when applying solid wood to an outside edge. The application itself isn't the issue; the problems occur when making the trim flush with the plywood surface. How do you get the edging perfectly flush without going too far and burning though the thin veneer? As hybrid woodworkers, we'll lean on the accuracy and control afforded by hand tools. Using a block plane, scrapers, sanding and a little blue tape, we can safely and accurately bring edge-banding flush to a plywood surface with minimal hassle and no regrets.

Making Edge-Banding

Stock for edge-banding starts as a solid board. Joint and plane the board so that it's a bit thicker than the plywood. The actual number isn't critical as long as it's thicker. You'll have to remove this excess material in a later step, so you don't want to be more than 1/16" oversized. Because plywood is typically thinner than its stated thickness, milling your stock to exactly ¾" or perhaps a bit proud of that usually gives you just the right amount. Then rip the board into ¼" strips at the table saw. If you don't have a sacrificial push shoe for this purpose, consider buying or making one. Cutting thin strips at the table saw can be dangerous without something to guide the workpiece completely past the blade. There are great commercially available guides; an alternative would be to cut the strips off the far side of the workpiece using a shop-made thin-strip jig.

Get a Grip ▪ The Micro Jig GRR-Ripper is an excellent tool for ripping thin pieces at the table saw.

Attaching Edge-Banding

There are a few schools of thought concerning the best way to attach edge-banding. Some folks like to use brad nails to get the job done quickly, but I am not a big fan of filling nail holes. Also, brad nails don't apply enough pressure to close up all of the little gaps between the edge-banding and the plywood, and a hairline gap nearly always results So to eliminate gaps, you'll need to use clamps and cauls. Cauls are lengths of wood placed between the clamp and the workpiece to distribute clamping pressure. As a caul for edge-banding, I use a long plywood cut-off that spans the entire length. Because the caul helps distribute the pressure, you can get away with fewer clamps than you might otherwise need. While several clamp types could be used for this task, I prefer parallel clamps. Not only do they feature a wide clamping head, they also provide a consistent platform for the workpiece to rest upon.

Edge-banding typically happens before assembly, when it's easy. Before applying glue, run a piece of blue masking tape along the face of the plywood on each side, right up to the edge. The tape helps catch any glue squeeze-out and assists in other ways during the flushing process. Apply a thin layer of glue to both the edge-banding and the plywood edge, then place the entire assembly flat on a series of parallel clamps with cauls in place. When you tighten

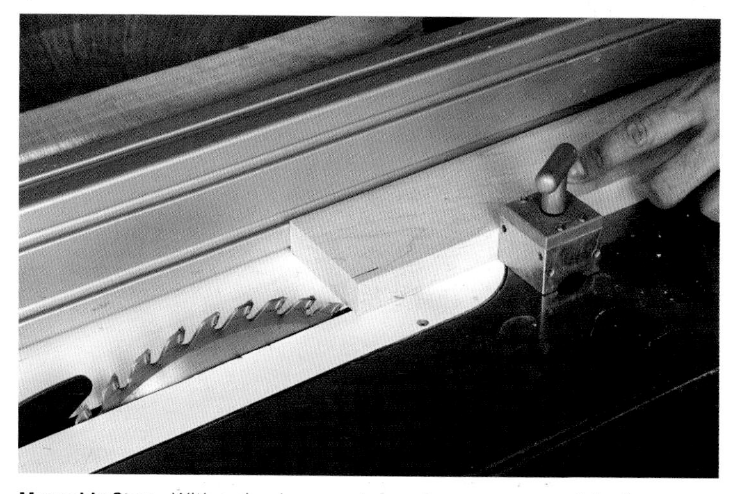

Moveable Stop ▪ With a simple magnet as a stop, you can reset the fence after each cut to rip thin strips of uniform thickness.

Simple Solution ▪ To make thin strips from boards that aren't too long, you can use a simple jig made from scrap MDF. The jig carries the workpiece through the blade.

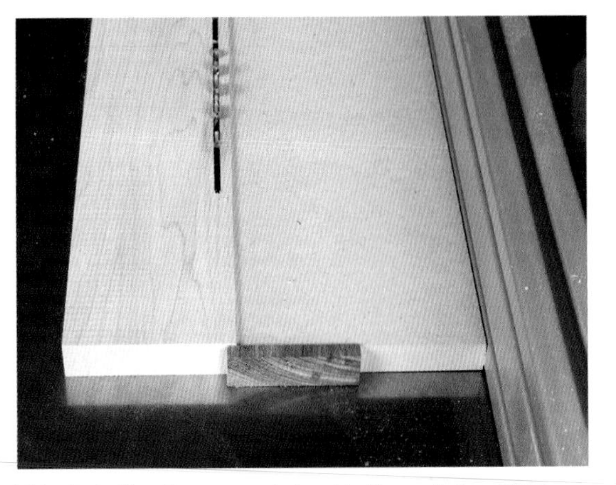

Thin-Strip Jig ▪ Scrap wood glued to the end of an MDF scrap makes a hook that guides the workpiece into the saw blade. With the jig between the rip fence and the blade, you set the fence only once, then rip all the strips you need.

the clamps, the slippery glue helps the edge-banding slide out of position and you'll need to make adjustments.

Use your fingers to make sure the edge-banding overhangs the plywood surface equally on both sides. Between the clamps being in the way and the glue squeeze-out blocking visibility, your fingers are the best tools for this job. If the edge-banding feels flush or below the plywood surface, loosen the clamp in that area, make a quick adjustment then reapply pressure. Once the glue skins over (30 minutes to two hours, depending on your climate), use a putty knife to scrape away the squeeze-out. With the blue tape in place, this should be a clean process.

Clamp Two Boards at Once

A cool way to eliminate the need for cauls while also making the best use of limited clamps is to clamp two pieces together at the same time. This is especially handy when edge-banding multiple shelves for something like a bookcase. Orient the two boards edge-band to edge-band and apply clamping pressure across both pieces. Each piece serves as a caul for the other. While clamping the boards this way does a fantastic job of spreading out the pressure, it does almost too good of a job. If the plywood edge isn't perfectly flat and features a slight dip here and there, gluing up this way may result in gaps between the edge-banding and the plywood. So for best results, place a layer of a soft material, such as cork, between the two pieces. The cork distributes the pressure and closes up any small gaps. Smaller cauls or one long thin caul also can provide localized pressure.

Efficient Clamping ▪ When gluing up edge-banding, it's a great idea to clamp two boards at once. Each board serves as a caul for the other and you need only a single set of clamps.

Trimming Edge-Banding Flush

Once the glue has dried, it's time to flush the edge-banding to the surface. If you are confident in your skills and perhaps a little daring, you can make quick work of this task with a block plane. The trouble is, if you go too far the plane will dig into the thin plywood veneer, which is bad. So I start with the block plane because every shaving saves me numerous passes with the next tool, either the card scraper or the No. 80 cabinet scraper. I keep the squeeze-out tape in place while planing, because it not only protects the surface, but also serves as an early warning system. The block plane has to go through the tape before it contacts wood, and it should be obvious when this happens because the tape folds over and sticks to the sole of the plane. That's how you know it's time to stop and move on to a scraper.

If you're using a card scraper, place a small piece of tape over the last inch of the side that overhangs the face of the plywood. You really don't want that side of the scraper to make any contact at all and a simple piece of tape will help prevent errant material removal as well as scratches and gouges. If you're using a cabinet scraper, set the blade for a fine cut. Carefully focus pressure on the edge-banding and take as many passes as you need to bring it flush to the surface. When the scraper begins removing material from the plywood face, even if it's just dust, you know the edge-banding is flush. Finish with a light sanding.

If you do a little research on this topic, you'll soon discover several power-tool solutions for trimming edge-banding flush to the surface of plywood, including table-saw jigs and flush-trim router bits. If I were making a kitchen's worth of cabinets, it probably would be worth the time and effort to set up one of these solutions. But for one piece at a time, it's just as easy and a lot less risky to do it with hand tools.

Use a Block Plane ... Gently ▪ The block plane can make quick work of trimming edge-banding, but you must be careful not to plane through the thin surface veneer.

Early Warning ▪ The blue tape tears up to warn you that the blade is getting too close to the plywood surface.

Scraper Is Safer ▪ When using a card scraper to trim edge-banding, tape the blade's sharp corner to protect the plywood. It's slower than a block plane but less risky.

Hairline Gaps

When edge-banding plywood it is very common to end up with hairline gaps between the solid-wood edging and the plywood. This can be due to imperfections in the plywood and/or edge-banding, or simply the result of uneven clamping pressure. Regardless of the cause, you should know how to fix it.

There aren't many occasions when I pull out the can of wood putty, but this is one of them. The hairline gap isn't a structural concern and only affects appearances. That's when wood putty makes the most sense. I use water-based non-shrinking stainable putty made by Timbermate. Using a putty knife, I pack the putty into the crack, let it dry, then sand away the excess. The gap is filled and the eye can't see the repair.

Small but Noticeable ▪ A hairline gap will drive you nuts every time you look at it.

Wood Putty Filler ▪ Fill the hairline gap with an appropriately colored putty.

Disappearing Act ▪ After some light sanding, the gap disappears.

Mortise for Butt Hinge

While some woodworkers certainly are more talented and skilled than others, I don't think the gap is as wide as people think it is. What separates good work from bad work usually is just a heavy dose of patience and some extra attention to detail. One job where patience really pays off is the installation of hinges. Installing a hinge is not difficult, but because it isn't immediately visible to onlookers, a woodworker might not feel the installation is quite as important as, for example, cutting perfect through-dovetails. This is unfortunate because a properly installed hinge truly is a thing of beauty. The hinge sits perfectly flush with the surface in a recess that exactly matches the length, width and thickness of the hinge leaf. The result is a door that is free to swing with no restrictions or interference, and the action feels fluid and smooth. With a little practice, anyone can achieve these great results.

Select Good Hinges ▪ Not all hinges are created equal. Don't be afraid to spend a few bucks for the good stuff.

The strict power-tool solution for creating hinge mortises involves a router and a shop-made template, and as a result a few issues spring up. First, I don't necessarily want to make a template for every style and size of hinge I have to install. Hinges vary widely in size and shape and sometimes even hinges from the same lot may differ ever so slightly in dimension and squareness. One size certainly does not fit all.

A Near Perfect Butt Hinge ▪ A perfectly installed butt hinge can be a thing of beauty, though I really should have aligned the screws, too.

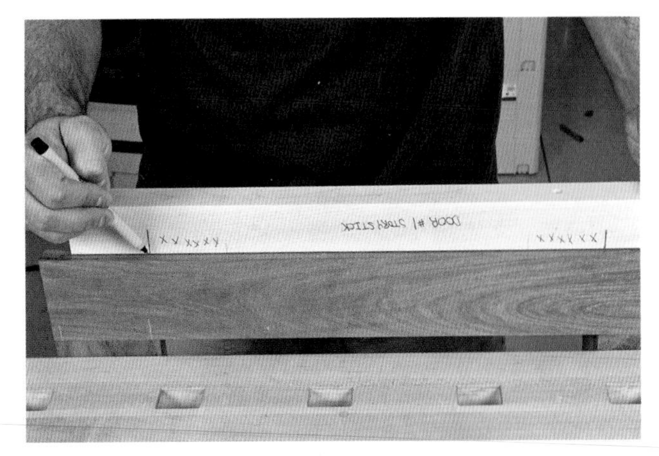

A Story Stick ▪ Use a story stick to lay out hinge locations on multiple doors. No tape measure needed.

Set the Cutting Gauge ▪ Use the hinge itself to set the cutting gauge.

Scribe the Mortise ▪ Use the cutting gauge to scribe the back edge of the hinge mortise.

Scribe the First Line ▪ Use a square and a marking knife to scribe the start end of the hinge.

The second issue is the risk of tear-out. By using a template, the router bit goes right up to the very edge of the mortise and if tear-out should occur, there is no buffer for repair. This is why I like to make mortises using the hybrid method. Instead of using the manufacturer's published hinge dimensions to make a one-size-fits-all template, we'll grab the information we need directly from the hinge itself and transfer that information to the workpiece. We'll then use the router for the bulk of the material removal and a chisel to finesse the mortise to an accurate scribe line.

Story Stick Aids Layout

We'll focus on the simple and fundamental butt hinge. The first step is to lay out the location of the hinge using a pencil. If there is more than one door, I usually create a story stick. A story stick is a thin piece of scrap wood with marks corresponding to the hinge locations. Referencing from the edge of the workpiece, the story stick allows for perfect repeatability on multiple doors without any measuring. For a butt-hinge installation, all we really need to do is mark the start and stop points of the hinge leaf mortise. How far back the mortise goes from the edge of the workpiece is information we'll extract from the hinge itself. Most butt hinges are happiest when the edges of the door and case are just short of the centerline of the of the hinge pin by $\frac{1}{32}$" to $\frac{1}{16}$". This establishes the point of rotation to ensure the door swings freely and gives a nice presentation when it's completely closed. By placing the marking gauge right against the hinge itself, we can quickly set it $\frac{1}{32}$" to $\frac{1}{16}$" shy of the distance between the end of the hinge leaf and the center of the hinge pin. With this measurement locked in, scribe the back of the mortise on the workpiece, from pencil line to pencil line.

Although we have pencil lines representing the start and stop points of the hinge, I like to kick up the accuracy by incising these lines with a knife. Use a square and a sharp knife to scribe along one of the pencil lines. With the knife resting gently in the cut, slide the hinge leaf into full contact with the knife blade. Press down to hold the hinge in place and remove the knife. Without letting the hinge shift, bring the knife to the other side of the hinge and make a slight cut. Don't scribe the full line because the hinge might not be square with the wood. Remove the hinge and drop the knife into the small cut you just made. Slide the square into contact with the marking knife and then scribe the full line.

Hinge Itself Gauges Depth

The only aspect of the mortise we haven't yet accounted for is its depth. I use a router and a small straight bit for the initial material removal, so it's important to set the depth accurately. Fortunately, we once again have no need of measurements. Using a plunge router, set the zero point by plunging down until the bit just makes contact with the bench surface. Now place the hinge leaf itself between the router's turret and its turret stop, and lock the turret stop in place. In theory, the router should now be set up to cut exactly to the thickness of the hinge. If you're using a fixed-base router for this

Locate the Second Line ▪ Use the hinge itself to locate the second line.

Scribe the Second Line ▪ With the knife in the starter notch, use a square to extend the cut line.

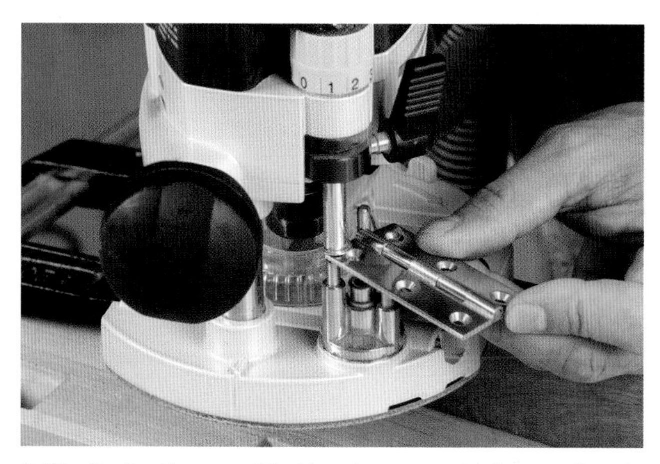

Set the Depth ▪ Use one of the hinge leaves as a depth gauge for setting the router turret stop.

Check the Depth Setting ▪ Test the depth setting by routing a test mortise in a piece of scrap wood. The goal is a flush fit.

operation, set the depth by turning the router upside-down and use the hinge itself to gauge the bit's position. Extend the bit until it is flush with the top surface of the hinge leaf. It's always a good idea to double-check your settings on a piece of scrap and adjust as necessary before you rout the good stuff.

Routing the Mortise

The routing process should be quick and the only real challenge is not going over the scribe lines. Get as close as you dare but there's no need to be a hero. With the bulk of the routing complete, you'll be able to fine-tune the mortise with a chisel.

In some cases you'll also face the challenge of balancing the router on a narrow workpiece, such as the edge of a door. This is unsafe and puts both you and your work at risk. To help support the router, clamp a piece of wood to the workpiece. In some situations you might want to clamp two pieces to the workpiece, one on each side. The more stable the router, the better the result. Just make sure the support pieces are perfectly flush with the top edge of the workpiece. When you're working on multiple parts at once, such as a pair of doors, it is often easiest to clamp the doors together. That way you not only have the additional support you need, but you can rout the hinges in both doors at the same time with only one setup.

Rout the Mortise ▪ Use a 1/4" straight bit and guide the router freehand to rough-cut the mortise to full depth but inside the layout lines.

Scrap Offers Support ▪ A piece of scrap wood offers additional support, making the router operation safer and more accurate.

Chisel the Ends First ▪ Use a sharp chisel to work back to the knife lines and clean up the end walls of the mortise.

Chisel the Back Last ▪ Because the small amount of wood at the back of the hinge is rather fragile, work this area last.

Clean up the Floor ▪ The mortise floor must be flat and smooth. Pare down any high spots.

Finesse with a Chisel

With the bulk of the mortise excavated, complete the job with a good sharp chisel. Because we are working with knife lines and not pencil lines, there is no ambiguity about where to place the edge of the chisel. Drop it into the knife cut and tap down. Start with the end grain first, making light taps with a small mallet until the chisel comes close to the bottom of the mortise. With the end grain chopped on both sides, it's safe to chop the back edge of the mortise without the risk of splitting the wood. With the chisel back flat on the mortise floor, use gentle taps or hand-pressure alone to remove any material that remains inside the scribe lines. To end up with clean and crisp corners, you might need to make a few more vertical chops along the mortise walls.

As an alternative, you could use the router plane to clean up the mortise floor but this usually isn't necessary. By the time you've set up the router plane, you could have already done the work with a chisel. And unlike other operations where the router plane helps finesse the final fit, that's not necessary here. With an accurately set-up router, the mortise depth should be perfect from the very start and cleanup can be done exclusively with a sharp chisel.

At first glance, the hybrid method may seem slower than the power-tool method and in the case of five or more hinges it just might be. But for the average project with two to four hinges,

Check the Fit ▪ Drop a hinge into the mortise, inspect the fit closely and adjust as needed.

Install the Hinge ▪ The hinge should rest snugly in the mortise, making it easy to drive the screws and finish the installation.

we could have the mortises cut and the hinges fitted in the same time it would take to make a mortising template. Keep in mind that even a template-routed mortise still needs to be squared up with a chisel, adding more time to the overall process. Even if the hybrid method happens to take a little longer (everyone works at a different pace), I wouldn't let that deter you from using these techniques. Something as critical as the action of a swinging door deserves your time and attention, and the hybrid method gives you the tools you need to install the hinges without a hitch.

Strength and Beauty ■ Dovetails are a wonderful detail to add to any drawer, box or case. Making dovetails doesn't have to be difficult.

Hybrid Dovetails

Hybrid dovetails provide the look, control and gratification of hand-cut dovetails but with a potential increase in speed and accuracy. Furthermore, because the band saw is doing the cutting, large dovetails can be cut rather easily using this method. In fact, the large dovetails between the front apron and end cap of my workbench were made using a similar method. With the help of a band saw, the cutting process is quick and easy, leaving all the fine-tuning work to our friendly neighborhood hand tools.

Dovetails are possibly the most revered, and sometimes feared, joint in woodworking. It might be an unpopular position to take, but I feel dovetails are overrated and the hoopla about making them is overblown. I can't argue with the fact that dovetails are useful and attractive joints, but they aren't the only game in town and you can spend too much time in pursuit of dovetail perfection, typically at the expense of something else. That said, you should understand how to make dovetails not only because they're useful, but also because the skills are transferable to other woodworking joints that are frequently used in furniture making.

Because of the way the joint is designed, the tails interlock with the pins, resulting in a joint that resists being pulled apart in one direction. This makes it a good choice for drawers where pulling on the handle places constant stress in one direction on the drawer face. If the glue should ever fail, the joint's structure will

Built to Resist ■ Something as simple as a drawer benefits greatly from the natural strength of a dovetail joint.

continue to resist the forces applied to it. Not only are dovetails exceptionally strong, they are also quite beautiful. Changing the arrangement, number and size of the pins can express everything from elegance to whimsy. It's no wonder modern woodworkers gravitate to the dovetail for both visual impact and strength.

Thanks to a growing interest in dovetails, we now have many choices in both tools and methods. A web search probably will find a few hundred blog posts and videos on the topic and most of them will be either variations of the classic hand-tool method, or router-based methods using jigs. The hand-tool method typically requires nothing more than layout tools, handsaws and chisels. Although the tools are indeed simple, the variations on the theme will boggle the mind. Questions you'll typically find yourself asking are, "Should I cut the pins or tails first? Do I use a pencil or a marking knife for layout? Is it better to use a Western saw or a Japanese saw? How wide should the pins be? What's the best dovetail angle? Do I need special angled chisels?" I can't answer these questions because they mostly boil down to personal preference, but I'll do my best to make sure you're prepared to make these decisions for yourself.

Condor Tails ▪ The massive dovetails featured on my split-top Roubo-style bench, called "condor tails" by Jameel Abraham of Benchcrafted, were easy to make using hybrid techniques.

The Traditional Tools ▪ Cutting dovetails by hand requires a few basic tools for layout, cutting and finessing the joints.

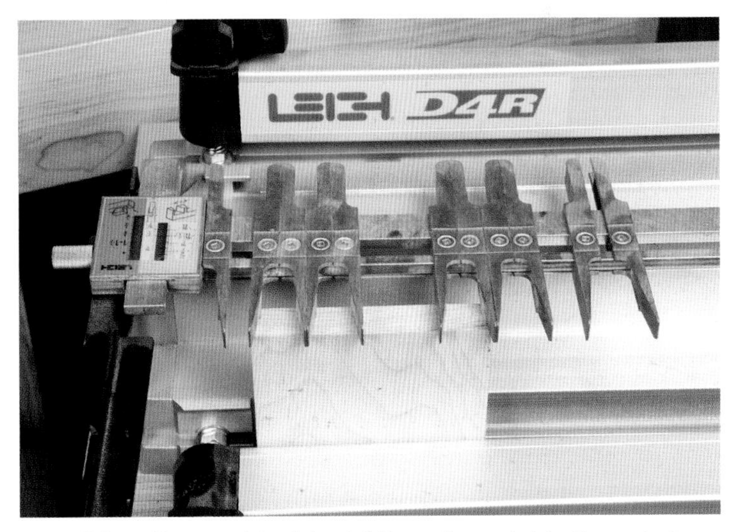

Dovetail Jigs ▪ There are lots of dovetail jigs on the market for those who have no desire to cut the joints by hand.

Dovetail Jigs

The dovetail jig is on the other side of the woodworking spectrum. Dovetail jigs use templates and router bits to cut joints quickly and accurately. Unfortunately, the best jigs are expensive and tricky to set up. Without a doubt I could have a drawer's worth of dovetails cut by hand and assembled in the time it would take to set up the dovetail jig. I have to reference the manual each time but once it is set up, the dovetail jig will knock out a whole mess of drawer parts in minutes. You really can't beat the speed and accuracy on larger projects.

There is one other drawback to using a dovetail jig that's worth mentioning, and that's the limit on pin size. One of the ways woodworkers like to show off their dovetail skills is by making narrow pins. When routing the pin

Pins and Tails

Many new woodworkers get confused about which board is the pin board and which one is the tail board. Many experienced woodworkers will tell you that the tail board is easily distinguishable because it looks like a bird's tail, which it does, but only when viewed from the face of the board. This can get confusing because the pin board, when viewed from the end, also looks like a bird's tails. So let's clarify this rule of thumb and say that the tail board is the one that looks like a bird's tail when viewed from the face. If you take a close look at the face of a pin board, you'll see that the pins look like little rectangles.

Which is Which? ■ The tail board (top) looks like a bird's tail when viewed from the face.

Another rule of thumb that might help you keep pins and tails straight is to think of them in the context of a drawer and simply memorize the fact that tail boards are always drawer sides. If a tail board was used for a drawer front, the joint could theoretically come apart from pulling on the drawer. The dovetail joint should be assembled in an orientation that resists the forces that will be applied to it. So for a drawer, that means the pin boards go in the front and back and the tail boards go on the sides. Early in my woodworking, I found that going through the exercise of disassembling a dovetail drawer in my mind was the quickest way to remember the differences between pin boards and tail boards.

sockets, the size of the router bit limits the size of the pins. Truth be told, small pins are just for show and do nothing to improve the strength of the joint – and if made too small, they can actually weaken the joint. But because they look amazing, I confess to making small pins myself. Hey, I'm only human.

Band Sawn Dovetails

One goal of the hybrid approach is to give the woodworker the best of both worlds, and making dovetails is no exception. While I like the small pins and the infinite pin and tail configurations offered by the hand-cut method, I also want the repeatability, predictability and speed offered by power-tool jigs. The band saw is a great compromise. By using the band saw to make the cuts in a predictable fashion at a fixed angle, instead of using a handsaw, you can reduce variability and decrease the amount of cleanup you'll do to arrive at a perfect joint. Additionally, the thinness of the band saw blade kerf allows you to create very small pins. If you happen to be good with a handsaw, the band saw method doesn't provide much advantage when doing only one joint. But when multiple joints are involved, the benefits become obvious: the band saw doesn't get tired. And for the average woodworker with average sawing skills, the hybrid alternative can help produce beautiful, tight-fitting dovetails with a hand-cut look each and every time.

Band sawing dovetails remains a hands-on method requiring chisel work and it isn't particularly fast, but it's a great alternative for folks whose handsawing skills might not be up to snuff or who suffer from chronic wrist pain. Cutting dovetails with a handsaw requires you to not only cut to a line, but also to cut straight while tilting the saw at an angle or rotating the workpiece in the vise so that the saw stays vertical. These cuts need to be repeated numerous times because the most modest dovetail series, for something like a small drawer, will require at least 12 saw-cuts per corner. If you're up to the challenge, handsawing skills are great to have in your back pocket, but they aren't exactly mandatory for the hybrid woodworker. I can make a pretty decent set of dovetails by hand, but I must admit I am slow – I can't expect to be proficient at a process I don't practice routinely. If your time in the shop is limited, I doubt you want to spend your precious shop hours making 100 practice cuts like the Karate Kid sanding the floor. By using the band saw instead of a handsaw, the weekend warrior can achieve dovetail nirvana while building skill at the same time.

Pin Layout

The band saw dovetail method follows the classic hand cutting method, so pins and tails still need to be laid out accurately. Some woodworkers like to use a marking knife and some prefer a set of dividers. I like to keep things simple so I lay out pins and tails using four basic tools: a pencil, an adjustable square, a cutting gauge and a bevel gauge.

The first step is to label the parts. Whether you're making a drawer, a box or a small cabinet carcase, you'll want to label the boards to indicate outside faces and adjoining corners. I usually give each corner a unique letter, writing that letter on both the outside faces and the edges. It may seem excessive, but thorough labeling is your only defense against dovetail disorientation.

Next, scribe the shoulders around both ends of the pin and tail boards, much like the procedure for scribing the shoulder of a tenon (page 122). The thickness of each workpiece dictates the setting for the adjoining piece. If the two pieces are the same thickness, the same setting can be used for both pin and tail boards. For a perfectly flush dovetail, the setting would be the exact thickness of the adjoining workpiece. I make dovetails to sit slightly proud after assembly, so I can plane away the proud material for a perfectly flush surface. If you aim for a perfect flush-fit, the pins and tails might come out slightly recessed thanks to cumulative error and you'll need to plane material from the entire face of each board to level the surface after assembly. That's more time and effort than I care to invest, so I set the marking gauge 1/32" more than the actual thickness of the adjoining board. The result will be pins and tails that sit 1/32" proud of the surface. Scribe the tail boards on both faces and both edges, but scribe the pin boards only on the faces. The pin boards will have half-pins at either edge so they don't require those scribe lines.

With one pin board positioned vertically in the workbench vise, use an adjustable square to lay out the pin positions on the end grain. The adjustable square is great for making symmetrical dovetail layouts because each setting can be used to make a mark from either side of the board. To help with aligning the cuts at the band saw, it's a good idea to mark the outside face of the board as well. This initial set of marks represents the pins at their thinnest point. To get a better idea of how the pins will

Label Clearly ■ When laying out dovetails, be sure to mark mating corners as well as which sides face out.

Scribe the Shoulders ■ A cutting gauge set about 1/32" more than the actual thickness of the adjoining board will result in slightly proud pins and tails.

Lay Out the Pins ▪ Mark the pins on the end grain and face grain.

Extend the Lines ▪ Set a bevel gauge to the desired angle (here a 1:8 slope) and extend the pin lines on the end grain.

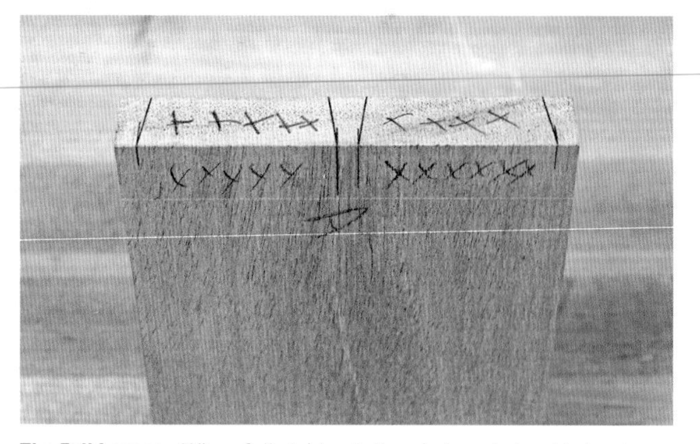

The Full Layout ▪ When fully laid out, the pin board should clearly show what wood to remove.

Tilt the Band Saw Table ▪ Use the bevel gauge to set the band saw table to the dovetail angle.

ultimately look, use a bevel gauge to extend the lines across the end grain. The general rule for dovetail angles is a 1:6 slope for softwoods and a 1:8 slope for hardwoods. I work almost exclusively in hardwoods so the 1:8 slope is my go-to angle. These angles are only general suggestions, so feel free to experiment with different angles to find something that looks good to your eye. Even if you have multiple sets of dovetails to cut, you only need to lay out one single pin board, assuming all of the parts are the same width and thickness. This is a big advantage over the traditional hand-cut method, which requires that each joint be laid out individually.

Band Sawing Pins

With the pin layout complete it's time to set up the band saw. Because the pins need to be cut at an angle (1:8 slope equals 7.1 degrees), the band saw table will need to be tilted in both directions. Most band saws have a table that tilts forward to 45 degrees, but has limited range tilting back. Fortunately, on most saws that limited range is enough to hit 7.1 degrees. The angle gauge and adjustment mechanism on nearly every band saw is inaccurate and shouldn't be trusted for fine work, so I use the previously set bevel gauge to get the band saw angle just right. With the band saw's guide/guard assembly raised and the bevel gauge against the blade, it's easy to find the perfect angle by observing how much light comes through the space between the bevel gauge and the saw blade. When the light is completely blocked out for the entire length of the bevel gauge, you know the table is at the correct angle.

Place the pin board against the band saw fence, outside face up, and observe the layout marks on the end grain. This will help you

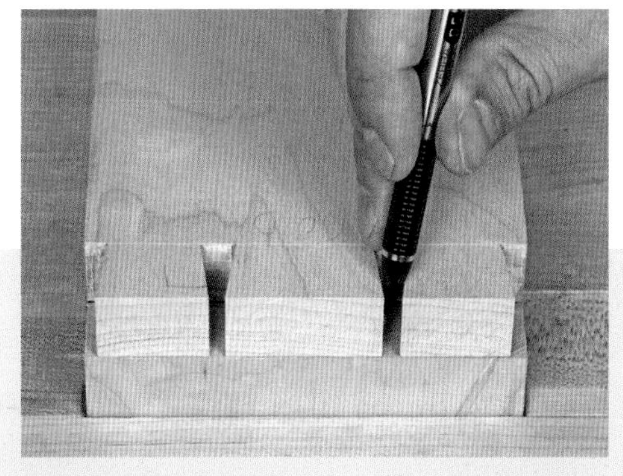

No Room to Work ▪ Pencils won't fit into tight pin sockets.

Pins or Tails First? Part 1

Everyone seems to have his or her own opinion about pins and tails and which to cut first. What makes this confusing to the new woodworker is the fact that there is no right answer; the dovetail gurus disagree. So how are you to determine which is the best method for you?

My advice is to think about the subsequent steps in the process. Let's say you want to flex your dovetailing muscles by making the pins really small. If you cut the tails first, you have to consider the difficulty of transferring the tail locations to the pin board. Because the pin sockets are so small, you won't be able to get a pencil in there; you'll have to use a marking knife, which might not be your preferred marking tool. For that reason alone you might be inclined to cut the pins first.

If you cut the pins first, you'll have a much easier time transferring the pin locations to the tail board, but you sacrifice simplicity in the sawing process at the band saw. To be continued in Pins or Tails Part 2 (page 165).

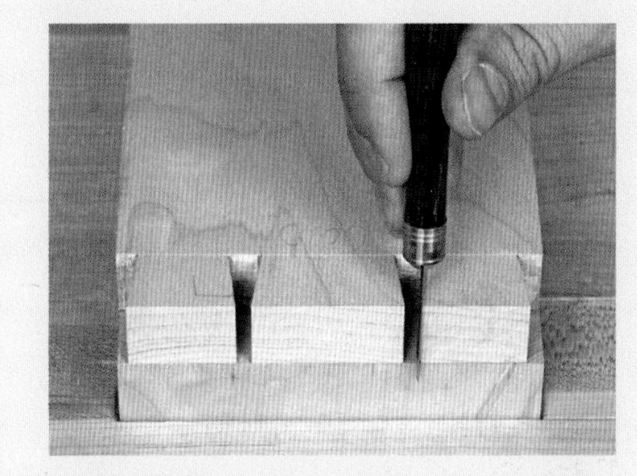

A Knife Will Fit ▪ Although a knife will fit, it might not be the best choice for layout.

visualize what cuts need to be made. Note that only half of the cuts can be made with the table tilted in this direction. Adjust the position of the fence for the first cut. It doesn't matter which cut you do first. I usually start at one edge and work my way across. As a rule, place the blade on the waste side of the cut and leave the pencil line intact. This is a good habit to get into, and its impact will be evident during the subsequent tail-cutting procedure.

When cutting at the band saw, take your time as you approach the shoulder scribe line. At all costs avoid sawing over the line, otherwise you'll be dealing with an unsightly divot in the closed joint. As the blade gets close to the scribe line, slow the feed rate down to a crawl. Most times the vibration of the saw alone will cause the blade to sneak up ever so slightly, helping you control the cut with great accuracy. When the cut is complete, gently pull the workpiece back while continuing to press it into the fence. The workpiece also will have pins on its other end, so

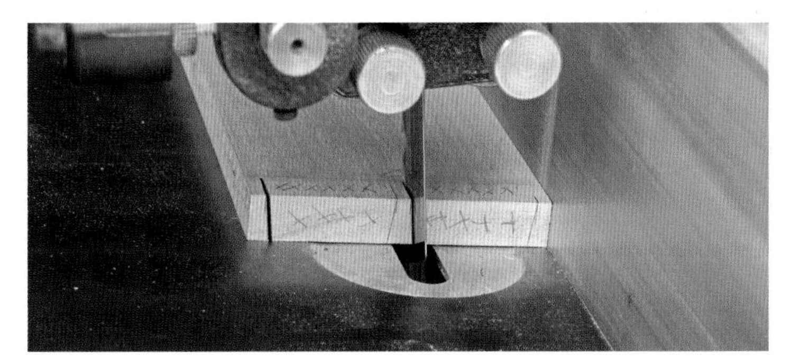

The First Series of Cuts ▪ Make the first series of cuts, keeping the band saw blade in the waste.

Watch the Shoulder ▪ You don't want the band saw blade to go beyond the shoulder line.

spin the piece around 180 degrees, keeping the outside face up, and make the same cut. Repeat these cuts on the rest of the pin boards, always keeping the outside face up. Maintaining this outside-up orientation ensures consistency and accuracy across the joints.

Reverse the Table ▪ Using the bevel gauge on the other side of the blade, reverse the tilt angle of the band saw table.

A Spacer Strip ▪ If your layout puts the fence too close to the blade, use a small strip of scrap as a spacer.

To adjust for the next cut, turn the saw off and wait for the blade to come to a complete stop. Adjust the fence and saw the next set of cuts on both ends of each pin board. When you have completed all the cuts that can be made with this setting, it's time to reverse the angle of the table. Once again using the bevel gauge for reference, tilt the table to the opposite angle. Make the second series of cuts in the exact same way as the first. The half-pin cuts might cause a conflict between the blade and the fence, so it might be necessary to use an auxiliary strip of wood to push the workpiece away from the fence by an inch or two.

With the pin walls established, tilt the band saw back to 90 degrees and carefully remove as much of the waste from between the pins as possible. Keep in mind that because the pins are sawn at an angle, it is all too easy to saw into their sides. This is why I always work with the skinny part of the pins facing down, allowing me to observe the widest part of the pin as I nibble away the extra wood. Once I have enough clearance for the band saw blade to make a turn, I run the blade parallel to the shoulder about $1/16$" away from the scribe line. It's far too risky and unpredictable to attempt hitting the scribe line perfectly, so the shoulder will have to be cleaned up later with a sharp chisel. The band saw won't be able to get everything so if space allows, I use a flush-trim saw at the workbench to cut away the remaining material. In tight spaces, a coping saw or fret saw also will do a fine job.

Clean Between ▪ The bulk of the material between the pins can be removed at the band saw.

Small Wedges ▪ Use a coping saw or fretsaw to remove the remaining small wedges of wood next to each pin.

Tail Layout

To lay out the tails, use the pin board itself as the template. Just as with the pin board layout, only one tail board needs to go through this layout process. With one of the tail boards sitting flat on the workbench with its outside face down, set the matching pin board up on end on top of the tail board, same as in the final joint. With everything set up properly, the thin part of the pins should be facing out. With the outside edges of the two boards nice and flush and the inside of the pin board aligned with the shoulder scribe line on the tail board, transfer the pin locations directly to the tail board. A sharp pencil is my preferred tool. I know what you're thinking: "But Marc, didn't you just tell us about using pencils for good joints and blades for great joints?" I did indeed, but this is a special case.

Much like the rule of keeping the saw blade on the waste side of the cut, always keep the bevel of the marking knife toward the waste side. The bevel is what gives the knife mark a width of its own, however small, and in this situation it would be doing damage to the tail material you want to keep. The entire space between the pins is keeper material and must remain fully intact for a snug, flawless dovetail. So as tiny as that knife line may seem, it will cut into the tail. If you use a pencil instead, you can mark the outer borders of the tails and set the band saw to cut in the waste just outside those pencil lines. Unlike a knife mark, a pencil line erases after the fact and doesn't do any damage to the wood.

For smaller parts, I do this layout process while holding the pieces together with my hand. There's no need for a more complicated setup. But for larger workpieces, it might be helpful to clamp the wood to the workbench while also using a large square block of scrap to aid in alignment.

Transfer Pin Locations ▪ With the pin board resting vertically on the tail board, transfer the pin locations. No additional support is required for small boards.

A Fine Line ▪ A .5mm mechanical pencil produces an accurate layout line.

For Large Boards ▪ When working with large boards, use a square piece of stock to provide support as well as a reference surface for alignment.

Band Sawing Tails

With the tails fully laid out, we can saw them at the band saw. I use a small angle fixture to hold the tail board at the appropriate fixed angle. This angle needs to match the one used previously to

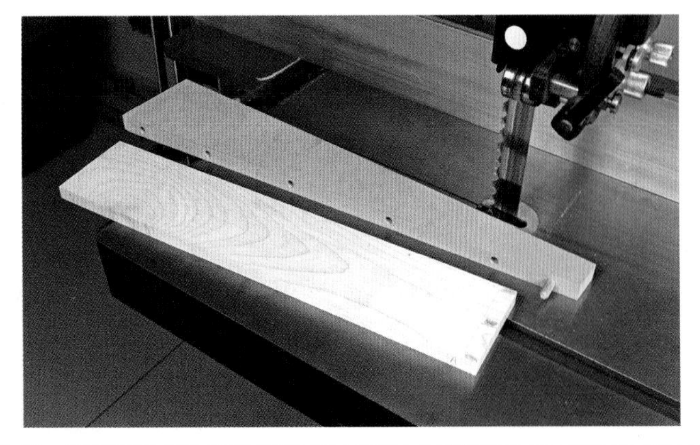

The Angle Fixture ▪ This MDF angle fixture will guide the tail board at the dovetail angle. The dowel stops the workpiece.

The Blade is in the Waste ▪ Set the blade square to the table and as always, line up each cut so the blade is in the waste.

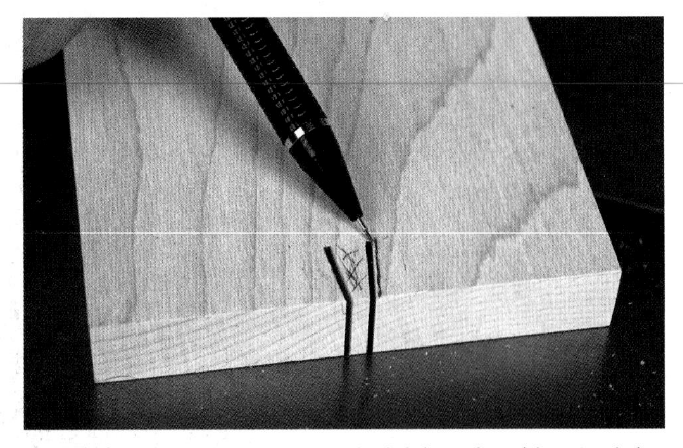

The Details ▪ The band saw cut on the left is perfect, it's not only in the waste but also leaves the pencil line. The cut on the right left too much material.

Reverse the Jig ▪ Reverse the jig to make the next series of tail cuts at the opposite angle.

cut the pins. Make the fixture from ¾" plywood or MDF. Saw the angle fixture straight and square on one side, with the desired angle on the other. Because the bevel gauge is already set to the exact angle, use it to transfer the angle to the fixture. Saw carefully to the line using the band saw and then smooth the rough edge with a quick pass over the jointer. The straight and square side of the fixture will ride against the band saw fence, with the workpiece up against the angled side. The workpiece must be immobilized by a stop to prevent it from sliding forward during the cut. The stop needs to be removable so that the fixture can be flipped end-for-end to make the second round of cuts; a dowel in a hole works perfectly. Drill a series of holes for a ¼" dowel across the angled edge. You'll most often use the outermost holes but the jig becomes more versatile with additional stop positions.

Insert a dowel into the leading or trailing hole (whichever you prefer) and slide the workpiece along the angled edge until it contacts the dowel. Now the entire assembly should be able to slide along the fence as one unit, which is critical to this process. With the tail board against the fixture and oriented with its inside face up, position the fence for the first cut, keeping the blade in the waste and leaving the pencil lines intact. If you're new to this process, err on the side of caution; there will be an opportunity to finesse the fit later. Just keep in mind that any material left outside the pencil lines eventually will need to be removed. If you are confident in your blade positioning and as long as you leave the pencil line intact, you can saw as close as you dare. With a little practice, you'll soon achieve perfect-fitting dovetails right off the saw with no finesse work required.

Once set up, you can make a cut on both sides of each tail board, keeping the inside face up. Once you've made all of the cuts that correspond to that particular angle, flip the fixture end-for-end. Flipping the fixture gives you the exact opposite cutting angle. Move the dowel to a new hole if necessary and proceed with the remaining cuts, adjusting the fence for each set. Use the band saw to remove the bulk of the material between the tails as well as on the outside shoulders. Just as with the pin board, leave about 1⁄16" of wood in front of the scribed shoulder line. Thankfully, the walls of the tails are not angled in the vertical dimension so it's easy to avoid accidentally sawing into the tails. If the pin sockets are small, you'll have to make repeated cuts straight into the socket. Be careful because it's easy to plunge too far and accidentally saw across the shoulder line.

Chisel Cleanup

Now it's time for cleanup. Both the pin and tail boards have rough, band sawn shoulders. Fortunately, the scribe lines are still intact and if you're good with a chisel, simply clamp the workpiece to the workbench with a sacrificial board underneath, and begin chopping. I recommend going only about halfway through the thickness of the board. Because you're working freehand, there's a good possibility that you aren't chopping perfectly vertical, so if you punch through to the other side you may end up cutting past the shoulder line. So once you have chiseled the entire shoulder about halfway

Tail Cuts Complete ▪ The tail cuts should look like this when complete.

Remove the Waste ▪ Use the band saw to nibble away the waste between the tails, being careful to not cut into the pins or the shoulder.

The Outside Shoulder ▪ Don't forget to remove the waste on the outside shoulder.

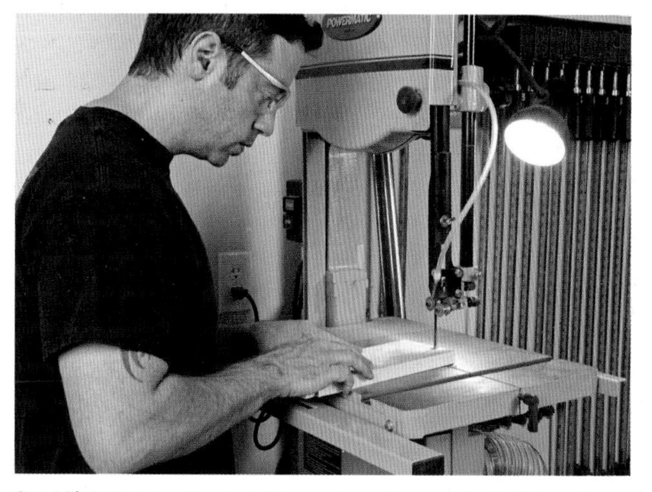

Good Light Helps ▪ Much of this work requires good visibility and a task light helps.

Chisel the Shoulders ▪ The back shoulder of the dovetails needs to be pared back to the scribe line.

Guide Block Help ▪ A thick hardwood guide block keeps the chisel square and prevents over-cutting.

Clean Up the Tail Boards ▪ Use the guide block for cleaning up the outside shoulders of the tail boards.

Watch the Scribe Lines ▪ Carefully watch all three scribe lines as you pare away the waste.

through, flip the board and begin chopping from the other side. It is likely that the area in the middle will need some additional work, so with the piece mounted vertically in the bench vise, use paring strokes to relieve any offending material and check your progress with a square.

I am skilled with a chisel, but I still prefer to do this shoulder work with the aid of a chisel guide. The guide is nothing more than a block of hardwood milled square that helps keep the chisel perfectly vertical. Align the block directly to the scribe line and clamp it in place. Not only does the guide keep the chisel on track, it also saves time. Because the chisel can't go off course, you can comfortably chop all the way through. There's no need to chop halfway and flip the workpiece for the second round.

With the help of the guide, the pin boards can be cleaned up in mere minutes. The tail board usually requires more care. Not only do you have outside shoulders to contend with, you also have tight pin sockets. The typical dovetail series features a half-pin on either side, so the tail board features two open half-sockets. The bulk of this material was cut at the band saw, but much like as on the rest of the shoulder, you'll need to use the scribe lines to guide the chisel and clean up to the line. If you choose not to use a guide block, you'll need to make a few light chops on all three sides of the outside shoulder. With clean shoulder lines established, position the tail piece vertically in the bench vise and pare away the remaining material. Because the corners will be visible in the final joint, it pays to be very careful here. Fortunately, if you choose to use a guide block, none of this shoulder work is necessary. Simply drop the chisel in place and chop through to the other side. Working the outside shoulder is no different than working any other part of the shoulder.

Depending on the size of the tail board's pin sockets, cleanup between tails could be tricky. If you push the limits with teeny tiny pins, you'll not only need a narrow chisel but you might also benefit from dovetail and fishtail chisels. The dovetail chisel features sharply beveled edges. The fishtail chisel has a skinny shank and a nearly flat blade that fans out near the tip. These special-purpose tools do an excellent job of cleaning up the internal angled walls of any pin socket, narrow or wide. But when it comes to extremely narrow sockets, a dovetail chisel is pretty much a necessity. My pins are typically small enough to cause minor cleanup issues when using standard bench chisels, but not so small to motivate me to buy a specialized set of dovetail chisels.

Test-Fit and Finessing

Time for the all-important test-fit. Lay one of the tail boards flat on the workbench and try dropping both pin boards into place, remembering that each joint has its own defined pin/tail pair and each should be evaluated individually. If you are skilled with aligning the band saw blade for each cut, you'll find that your pins and tails nest together perfectly. If that's the case, congratulations. If that isn't the case, and you followed these instructions closely, you most likely have a tight fit because you erred on the side of caution by leaving a little extra material. You can always remove wood, but you

Pins or Tails First? Part 2

At the risk of confusing my woodworking comrades, I need to highlight another factor to consider in your quest to determine which is best: pins first or tails first. If you were to cut the tails first instead of the pins, you could save yourself quite a bit of time and effort on the tail cuts by simply flipping the workpiece. If the tails are symmetrical, every cut has a mirror-image that can be sawn with the fixture in the same position by simply flipping the workpiece. Every time we adjust the fence, we introduce error, so why the heck aren't we using this flip trick even though we cut the pins first? The reason is, we have no guarantee that the pins are perfectly symmetrical.

Potential error always exists in marking and cutting the pins. This error is nullified by using the pin board itself to transfer the pin locations to the tail board. Even if something is off, it won't affect the result as long as we saw to our pencil lines. So at this stage, if we assume that the pins are perfectly symmetrical and employ the flip trick, we could be in for a rude awakening.

If we were to cut the tails first we could safely employ the flip trick, resulting in perfectly symmetrical tails. Those tail locations would then be transferred to the pin board and we'd be off to the races, right? Not necessarily. Remember the small pin issue we discussed in "Pins or Tails First Part 1?" Not being able to fit a pencil in the pin sockets is a deal-breaker, so we have no choice but to go pins first for this operation. If you like your pins a little wider, you might consider tails first as the better option. See, I told you the choices might boggle the mind.

Cut and Flip ▪ When cutting tails first, you can reduce setup by first making one cut, then flipping the workpiece over to make the second cut.

can't put it back. So if your joint happens to be a bit loose, take a close look at the tails. You will probably notice that the pencil lines are gone. If you leave the pencil lines intact and always cut in the waste, it's impossible to have a loose fit. If you accidentally over-cut, don't be discouraged. It takes a few tries to get the feel for hybrid dovetails and until you gain confidence at the band saw, the pins probably won't glide perfectly into their sockets. The good thing is that you can easily identify what went wrong. The only mistake worth getting frustrated over is the one where you can't determine why or how it happened. If you know the whys and hows, you can avoid the mistake in the future.

If the fit is tight, inspect the tails to determine where there might be extra material. Most times, it is clearly visible with close inspection. Use the chisel to pare away the extra material on the offending tails. To make sure you are paring the tails uniformly, try covering the surface with pencil marks. When all of the pencil has been removed, you know you worked the surface evenly. You can continue adding pencil marks and paring them away until you have removed the appropriate amount of wood.

If you can't quite determine the source of the tight fit by visual inspection, you'll have to do more investigative work. Start by applying a small amount of pressure to the pin board while rocking it back and forth over the pin sockets but don't force the joint to close. Take a close look, you should be able to identify which tails are too fat and which ones are likely to slide right in if there was no interference. With the tail board in the bench vise, take light paring strokes to remove the offending material. Check your progress by doing a test-fit early and often. It might take several rounds before you get the joint to go together without the aid of a sledge hammer, but patience will pay off. And keep in mind, wood does compress so don't be too forceful with the test-fits.

Assembly ▪ Test-fit each joint separately. If all goes well, dry assemble all four pieces.

Inspect ▪ Check the entire structure for square and inspect each joint to see if there is room for improvement.

Too tight? Pare away! ▪ If you find a joint is a little too tight, mark the offending tail with pencil and lightly pare away some wood.

Once the joint goes together by hand pressure or with a few light taps from a dead-blow hammer, you're good to go. Note that the pins and tails are sitting proud of the surface as intended. After final assembly when the glue has completely dried, use a block plane to level the pins and tails to the surface. Follow up with a few passes from a random-orbit sander or a scraper. Depending on the project and your tastes, you may want to work the surface a bit more to remove the shoulder scribe lines. Some people like to keep them because they are a clear and obvious indicator of handcrafted dovetails. Others find them unattractive and are happy to remove them. The choice is yours.

By the time you band saw your second or third set of dovetails, you should be producing nearly perfect results right off the saw. If you're having trouble lining up the blade, try shedding some light on the subject. A small task light on the outfeed side of the saw will greatly increase visibility in an area that is typically plagued with shadows. If you have a head lamp, that could also be quite helpful.

Dovetail Assembly

For the most part, dovetails are a self-squaring joint. If the fit is snug and each piece nests within the other, the joint will usually end up square. But the woodworking gods aren't always benevolent and it's important to make sure that your glue-ups are square and the joints are closed tight. Once each joint has been finessed and fits acceptably, add glue to the pins and tails and apply two clamps in each direction. With the joints all closed up, use a square to evaluate each corner. If the assembly is out of square, throw a fifth clamp across the assembly from corner to corner. By applying a light amount of pressure, you can adjust the squareness to perfection. If you happen to over-correct, you may need to move the clamp to the other two corners to bring it back to square.

Squareness isn't your only concern during assembly. You also need to make sure each joint has adequate pressure, otherwise unsightly gaps will result. If you run the clamps right

Assembly ▪ Each joint receives glue on all mating surfaces, with four clamps to apply pressure.

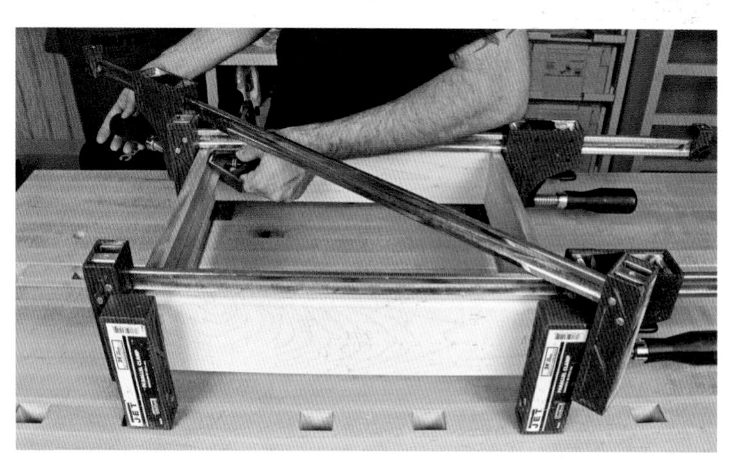

Check for Square ▪ Dovetail joints are mostly self-squaring but it's always good to check all four corners. Apply a diagonal clamp to make adjustments.

Small Scrap Focuses Pressure ▪ It may seem like overkill, but I glue small pieces of scrap to the surface to apply pressure in just the right places.

Trim the Pins ▪ The proud pins can easily be trimmed flush to the surface with a few strokes of a block plane.

Trim the Tails ▪ The proud tails can be trimmed flush to the surface in the same way as the pins.

over each joint, you'll never get the pressure you need, because you kept the pins and tails proud and they are going to prevent you from applying pressure directly over the joint. The easiest way I know to get around this problem is to apply small pieces of scrap wood to the faces of the pins and tails, using a tiny dab of CA glue to hold the pieces in place. When the clamps are tightened up, they engage with the scrap blocks instead of the proud pins and tails, applying pressure exactly where it's most useful and resulting in a tightly closed joint. The small pieces of wood are easily removed or planed away after the glue dries.

Because it can be tedious to find and attach small pieces of scrap wood to aid in clamping, I should mention that there is another strategy. Instead of having proud pins and tails at this stage, you can employ the opposite strategy, making your pins and tails slightly recessed. To do this, set your cutting gauge 1/32" less than the thickness of the adjoining workpiece, exactly the opposite of what we did in this example. The result is a surface that requires no additional clamping aids. Also, with this approach you can fit the drawer fronts and backs to their openings before cutting any dovetails, removing uncertainty from that part of the process. The drawback is, after the glue dries you will have to plane all four faces of your drawer or box down to the end-grain of the joint. This isn't so bad if you only have one drawer or a small case. If you have many drawers or a large piece of casework, that could be a heck of a lot of additional planing. So pick your poison. Most times, I find it easier to keep the pins and tails proud as mentioned previously.

When making dovetails with thicker stock (3/4" or more), it is often possible to place your clamps just adjacent to the pins and tails, negating the need for any clamping aids at all. Because the stock is thicker, it is less prone to bending under clamping pressure, so clamping next to the joint usually provides adequate pressure.

In my opinion, this system offers the best of both worlds: the look of hand-cut dovetails with the increased speed and accuracy of the band saw. I still use my router and dovetail jig on large projects with numerous drawers, but on smaller pieces with five drawers or fewer, the hybrid band saw method is an excellent choice. The fact that this system is also useful for large dovetails makes it even more valuable. When it's all said and done, you'll have a set of strong and elegant dovetails that will stand up to scrutiny by the most discerning of dovetail aficionados.

Curves

Most woodworkers begin their journey with basic straight-line pieces. The simple beauty of the Craftsman and Mission styles draw many new woodworkers into the fold. For some, this type of furniture is a stepping stone for their evolution into work that includes curved elements. This might not sound like too big of a deal but when it comes to the actual woodworking process, curves can throw all sorts of curve balls at the designer/builder, because good joints rely on flat and square reference surfaces. This is why you'll find most joints being cut ahead of the curved features. Incorporating curves into your work is a deep topic and what we are going to discuss here is just one aspect: creating the curve itself.

To make curves with machines, draw the shape on the workpiece and cut it out with the band saw or jig saw. The resulting cut always requires additional shaping and smoothing, and sanding is typically the solution. Oscillating spindle sanders and bladder sanders are often the tools of choice and they can be effective if the curve happens to match the sander's profile. But in most cases, project curves are wide, gradual and subtle. Spindles and bladders are only a few inches in diameter and as a result, they tend to create minor hills and valleys in the curved surface.

One alternative is to create a template and use a router with a flush-trim bit to cut an exact copy of the template profile. This is a quick and efficient way to create perfect curves, but in order for it to work the template curve needs to be perfect. All the issues of creating a curved workpiece are also present when creating a template. Minor imperfections in the template will be transferred to the workpiece.

More work is needed for a truly smooth and continuous curve, whether on a template or a workpiece, and the hybrid approach provides an excellent solution. Instead of using a machine to fair the curve, we'll rely upon a spokeshave, a rasp and a flexible sanding strip. These tools give you an incredible amount of control while helping to eliminate hills and valleys. The result will be a smooth and continuous curve. When making multiples of a curved piece, you'll do this fairing process only once, on the template itself. Then you can use the template to cut all the copies you need. Let's review the hybrid curve-cutting process.

To cut a curve, you first need to draw the curve. Not being terribly artistic myself, I rely on several tools to help draw elegant and graceful curves that are pleasing to the eye and serve the overall design of the project.

Curves Can Be Fun ▪ This small table of my own design features playful curves and an organic form.

French Curves are Handy ▪ French curves can help you draw just about any curve you could imagine.

Bender Boards ▪ Large symmetrical curves can be made with a thin piece of scrap and a few nails.

Drawing Bows ▪ Lee Valley Tools sells symmetrical and asymmetrical drawing bows, useful when drafting curved pieces.

French Curves

For small-scale work with asymmetric curves, I use plastic French curve templates. Most of the time, I have two or more points that need to be connected with a particular curve that I can visualize. The problem is, I can't draw it freehand. Thankfully, the French curve gives me numerous options for getting from point A to point B in ways that can match the idea I already have in my head.

Bender Board

For symmetrical curves on any scale, try a bender board. It's just a bendable strip of thin man-made material. While you can certainly use a hardwood ripping for this purpose, I find that MDF and plywood give the most consistent results. By securing the bender board with clamps, nails or weights, you can create just about any symmetrical curve. If you have a second set of hands in the shop, you can skip the clamps, nails or weights, and ask the other person to bend the strip by hand while you trace the shape with a pencil.

Drawing Bows

Lee Valley Tools sells a set of fiberglass drawing bows, one for symmetrical curves and the other for asymmetrical curves. Both bows feature a locking strap that allows you to adjust the radius of the curve and lock it in position for tracing. This is a huge help in a one-man shop and that feature alone makes it worth the price. On the asymmetric version the bow is tapered in thickness, so when you tighten the strap, the thinner side bends more than the thicker side. The result is a pleasing asymmetric curve. This type of drawing bow would be fairly easy to make in an afternoon, after a trip to the craft supply store.

Shaping the Curve

However you draw the curve, the next step is to cut the shape using the band saw or a jigsaw. Get as close to the line as possible without crossing over, though if you're new to this process, stay 1/16" outside the line. The good news is that even if you do go too far, curves are forgiving and usually you can make a quick change on the fly. After sawing away the waste, head to the workbench to fair the curve. The goal is to work back to the pencil line, creating a smooth and continuous shape. I use several tools including the spokeshave, the rasp and a flexible sanding strip.

Spokeshave

Start with the spokeshave. The band saw leaves a rough washboard surface and if you stayed 1⁄16" away from the pencil line, there's going to be a good amount of material to remove. The spokeshave's small sole does efficient work on both tight and wide curves. For inside curves, use the convex spokeshave. For outside curves, use the flat spokeshave.

The rule with the spokeshave is always to work downhill with respect to the wood grain. When the curve changes direction or the grain becomes temperamental, simply reverse the tool and push instead of pulling. You can flip the workpiece around if you're only comfortable pushing or pulling and you want to keep your body motion consistent. Becoming comfortable both pushing and pulling a spokeshave speeds up the work and reduces the time spent repositioning the workpiece.

Continue to work to the pencil line, and be sure to check the edge for square. It's easy to accidentally put more pressure on one side of the spokeshave than the other, taking the edge of the workpiece out of square. If you can identify problem areas before they get out of control, they are easy to fix by adjusting your hand pressure.

Rough Out the Curve ▪ At the band saw, cut the curve roughly to shape.

Don't Saw the Line ▪ Try to stay at least 1⁄16" away from the line during the entire cut.

Traditional Shaping ▪ Spokeshaves are great for sculpting and refining curved surfaces.

The Versatile Spokeshave ▪ The small size of the spokeshave means it can go into places other hand tools can't.

A Rasp Sculpts, Too ■ A rasp is a great tool for finessing curves and helps to create a smooth and consistent surface.

Trust your Fingers ■ Your fingers are a very sensitive tool; you can feel when a curve is smooth enough.

Finish with Sanding ■ Finish curves with a bit of sanding using a flexible sanding strip (page 174).

Even if you eventually intend to round over the edge, keep things nice and square until you've fully established the curved shape.

Rasp

The spokeshave's narrow sole follows any hills and valleys that are wider than the tool. That's why the wood rasp is the second tool in my lineup. The goal with the rasp is to remove any high spots, much like a handplane. The rasp leaves behind a characteristic texture of scratches, so it's easy to see the high and low spots after just a few passes. Marking the surface with pencil lines may also help.

For safety's sake be sure there is a handle on your rasp; without one, if the rasp was to catch, you could drive its tang into your hand. Hold the rasp on the surface with its flat side down (most rasps have a flat side and a convex side), and close to parallel to the edge. The rasp can't be held perfectly parallel to the edge except when working on a tight-radius outside curve, so it's usually held between 15 degrees and 45 degrees off parallel. Even with this angled orientation, the rasp can cover a great deal of territory with each pass and it quickly knocks down any remaining high spots. This is a crucial step in the fairing process because you are no longer concerned with the pencil line. Instead, the goal is a flat, smooth continuous curve with no major hills or valleys.

Your eyes and fingers are your two best tools for gauging progress. If the curve doesn't look right, you'll spot it right away. But there's always a point where your eyes can no longer help and the fingers step in for the final inspection. I usually close my eyes and slowly rub my fingers across the curved surface. No, this is not a hippy-dippy way of becoming one with the wood. Your fingers can detect surface irregularities down to thousandths of an inch. When running your fingers along a curve, you'll be able to detect hills and valleys that the eye simply can't see. When the surface feels smooth and continuous, it is.

Sanding Strip

To finish off the curve, I use a flexible sanding strip, which is nothing more than a thin strip of wood with sandpaper attached to it. I have several of these strips outfitted with different grits, usually #120, #180 and #220. The bendable strip conforms to the shape of just about any inside or outside curve, but because it is fairly rigid, it doesn't follow the hills and valleys. This tool allows you to put the finishing touches on a curve while producing a finely sanded edge.

Templates

When I need to make multiple copies of a curve, the first completed curve can serve as a template. Use a pencil to trace the completed curve onto the new workpiece, cut out the shape and repeat the fairing process. Or better yet, use double-stick tape to secure the two pieces together and use a router with a flush-trim bit to make a perfect copy of the completed curve. The routed surface probably will require some smoothing, but the flexible sanding strip is usually all that is needed.

If your project requires numerous copies of a curve or perhaps you expect to build the same project again in the future, it's a great idea to make a template from ¼" plywood or MDF. The template can then be traced onto each workpiece, and it also could become the primary

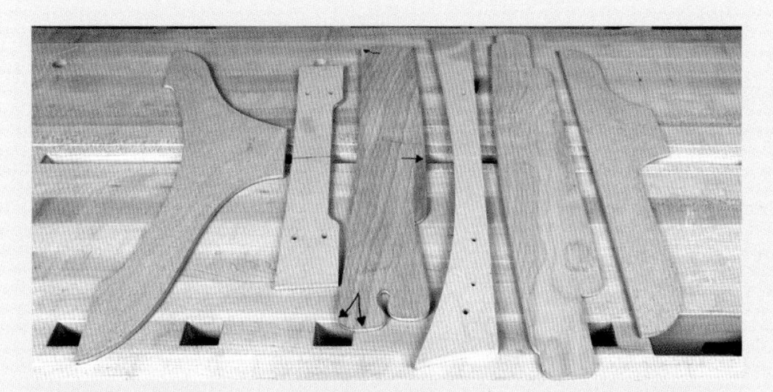

Templates Everywhere ■ Keep your project templates because you never know when you might need one again.

Double-Stick Tape ■ Double-stick tape is strong enough to attach the template to the workpiece.

Flush Trim ■ A flush-trim bit at the router table brings the workpiece into final shape, using the template as the guide.

A Perfect Copy ■ When it's all done, you're left with a perfect copy of the template's profile.

Flush-Trimming with Templates

Whenever you have multiple workpieces that need the same curve or profile, consider making a template. While it might seem like extra work, a template will save time and frustration. Use the template to transfer the shape of the profile to the workpiece blank and cut it to rough shape at the band saw. Be sure to stay 1/16" outside the line. Attach the template to the workpiece using double-stick tape. At the router table, carefully push the workpiece into the bit and let the bearing ride along the template. With only 1/16" of material to remove, the router bit should have no problem sculpting the workpiece into shape. Pry the template off with a putty knife and repeat the process on the next workpiece. The awesome part of this technique is the fact that you can make as many copies as you want and then store the template indefinitely for future use. Think of it as a functional historical record of your woodworking history.

router template. When you're finished, label the template with the project and the date and stow it away. Whenever I'm making a curve for a project, I always search through my template collection to see if an existing template will do the job. You'll be surprised how often those old templates come in handy. Making the template itself is no different than cutting the initial curve in a workpiece, except it's actually easier when the thickness is only 1⁄4".

Make a Sanding Strip

The key to a good flexible sanding strip is the strip material and its thickness. If the stock is too thick, it doesn't bend. If it's too thin, it is liable to break with use. I usually make sanding strips from solid wood, sliced to about 1⁄8" thick at the band saw. If you have a drum sander in the shop, you can use it to fine-tune the final thickness to your liking. Otherwise, smoothing the sawn strip with a random orbit sander should do the trick.

The strip should measure about 4 1⁄2" wide by 10" long, which is half the width of a standard sheet of sandpaper and an inch shy of its full length. Undersizing the length leaves extra material for securing the paper to the strip. Although optional, I find adding a layer of thin cork provides some cushion while also making the strip more durable. Cut a piece of thin cork to the exact dimensions of the strip and attach it with spray adhesive.

Cut a piece of sandpaper in half and put it on a flat surface, grit-side down. Place the wood strip on top and fold the protruding sandpaper up and over the ends. With the crease established, use spray adhesive or cyanoacrylate (CA) glue to secure the flaps semi-permanently to the top of the strip. When it comes time to replace the paper, you will be able to remove the adhesive with a scraper. Sometimes I find it helpful to put a small amount of adhesive on the pad under the main sanding surface; this could result in the cork tearing out during replacement, so be careful if you try it. The final touch is to add a simple set of handles. Use CA glue to secure the handles in place.

Lay Out the Parts ▪ The flexible sanding strip consists of sandpaper, cork, a thin strip of wood and two handles.

Attach the Cork ▪ Attach the cork to the wood strip using spray adhesive.

Glue the Handles ▪ The handles are glued on using CA glue.

Attach the Sandpaper ▪ Fit the sandpaper onto the strip. Fold the extra paper over the top and glue it down.

Just the Right Flex ▪ When the strip is done, you should easily be able to flex it with your hands.

Gossamer Shavings ▪ There's nothing quite like a wispy shaving coming off of a handplane. Instead of kicking microscopic particles of dust into the air (and into my lungs), I'm making shavings. This is one of the most beneficial, satisfying and gratifying changes a woodworker can make to his or her shop methodology.

Surface Preparation

Every project that leaves my shop is smoothed at one point or another with sandpaper. Sandpaper smooths and levels surfaces, removes mill marks, eases sharp edges and smooths finish between coats. But for all of the great things sandpaper does, most woodworkers consider sanding a necessary evil. If done properly it can take a very long time. For a board that comes right out of a typical straight-knife planer, I might start sanding at #80 grit to remove the mill marks and follow up with #120 grit, #180 grit, and sometimes #220 grit. Even with a random-orbit sander, this is extremely time-consuming and tedious. Stocking up on all the various grits and types of sandpaper is expensive, too, especially when you purchase quality sanding discs. And let's not forget about the loathsome dust generated by sanding.

One of the biggest hazards we face as woodworkers comes from the dusty air we breathe. Many people assume that if they can't see the dust, it isn't a problem. The reality is that the finest dust is the most dangerous because it can become lodged deep inside our lungs. I am fortunate to have a high quality dust extraction system attached to my sander, but many don't. And nothing can collect the dust generated by the inevitable hand sanding that is necessary on certain project parts and finishes. So the less dust we put into the air, the better.

With all this in mind, how can the hybrid woodworker reduce dependency on sandpaper? Fortunately, there are several tools in the hybrid kit that will help, including the card scraper, the cabinet scraper and the smoothing plane. If all you did was use a cabinet scraper to remove mill marks left by the planer, you could at least cut out #80-grit and #120-grit paper. If you fine-tune the cabinet scraper for a light cut and also use the card scraper, you could also remove #180-grit and even #220-grit paper from your shop. Of course if you are so inclined, there's nothing quite like the surface left behind by a well-tuned smoothing plane. If the board is nice and flat, several passes with a smoothing plane could be all that's needed before adding finish. I know I'll never get rid of sandpaper completely, but cutting out a few grits does the body good as well as the wallet.

In my shop smoothing generally depends on the project and the wood species, but here's a typical scenario. After cutting the joints, surface preparation begins. All powered cutting tools leave mill marks. The marks range from blade tracks to washboard patterns to burns. Even though on the macro level the surfaces are straight and flat, on the micro level they still need a bit of love. If the wood is not temperamental or prone to tear-out, I will reach for the smoothing plane. Nothing smooths the surface faster. Within just a few passes, the surface can go from rough to ready for finish. When it comes to the smoothing plane though, you really need to be confident that it isn't going to cause tear-out. At this stage of any project, you certainly don't want to mar the surface. Careful board selection and attention to grain orientation will stack the cards in your favor in this regard. But budget restraints and reality usually dictate that we make do with the boards we have, perhaps resulting in an imperfect and challenging scenario for planing. Fortunately, in

Sanding Isn't Much Fun ▪ While it's necessary, I don't find sanding much fun. Hybrid techniques dramatically reduce our need for sandpaper.

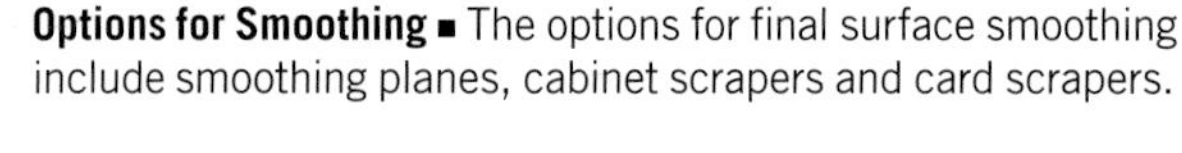

Options for Smoothing ▪ The options for final surface smoothing include smoothing planes, cabinet scrapers and card scrapers.

Smoothing with a Smoother ▪ A smooth plane is capable of putting a glass-smooth surface on the workpiece.

Mill Marks Just Disappear ▪ A No. 80 cabinet scraper is a great tool for removing mill marks left by the power tools.

those cases, we have scrapers. And scrapers don't care much about grain direction.

I usually begin by removing mill marks using the cabinet scraper. With its wide sole, this tool excels at knocking down high spots. My cabinet scrapers are tuned for an aggressive cut so the surface that's left behind is smooth, but not as smooth as it could be. This is where the card scrapers come in. With just a pass or two, I can remove any marks left by the cabinet scraper and leave the surface smooth. While some folks like the look and feel of a freshly scraped surface, I still prefer to give the wood a final sanding. Most times I use my random-orbit sander with #220 grit but you can also hand-sand with a sanding block. The surface doesn't need much work so the sanding time is short.

I can't get rid of sandpaper completely because I still rely on it for the final sanding of surfaces and detail work, as well as for sanding the finish, but I have significantly reduced my dependency. Instead of kicking microscopic particles of dust into the air (and into my lungs), I'm making shavings. This is one of the most beneficial, satisfying, and gratifying changes a woodworker can make to his or her shop methodology.

Card Scrapers ▪ Card scrapers can remove mill marks and, with a fine burr, prepare the surface for finish.

Sanding Still Happens ▪ After the hand tools work the surface, a light #220-grit sanding blends everything together.

Oops! ▪ Sanding this panel too aggressively resulted in a loose fit and an unsightly gap.

When to Smooth?

Knowing when to smooth a project board is key to your success. If you size a dado for a shelf and then smooth the shelf after cutting the dado, you may end up with a loose joint. But in most other cases, it's a good idea to smooth the boards after cutting the joints. Smoothing removes wood. Because some surfaces require more work than others, your project boards may end up at different thickness after the smoothing process. So if you decide to cut the joints after the smoothing process, your results may be less than ideal because a key component to good joinery is material milled consistently. So before smoothing any project part, consider its purpose and its destiny, and determine the best time to begin the process. And it is almost never smart to leave surface preparation until after assembling your project. The exception, of course, being parts that need to be flushed up, such as door frames and drawers.

4

Hybrid Woodworking Projects:

The Best of Both Worlds

Woodworking is like playing music. Once a song is written, a good musician doesn't have to stress much about each individual note or the instrument he uses to play them. Instead, he plays the notes as part of a cohesive whole. The same can be said for woodworking. Eventually, you won't have to stress about which tool is best for the job and what's the most effective way to use it. You'll just know and your instincts will drive the project to completion. The result will be a sweet wooden melody.

Best of Both Worlds ▪ One of these Shaker tables has more handwork in it than the other. I can tell the difference. Can you guess which table I like better?

While this state of woodworking zen is our goal, it will take time to get there. One of the best ways to progress on your hybrid woodworking journey is to observe how others get the job done. So to help you along, I thought it would be useful to show you a basic outline of five of my projects. For each project, I'll discuss a few notable areas where hybrid techniques were used. You'll recognize all of the techniques because they were discussed earlier in this book; seeing them in context should help you understand them better.

Each of these projects is available for purchase as a video course at www.thewoodwhispererguild.com.

Big and Beautiful ■ This platform bed made from bubinga, wenge and maple reflects the exotic tastes of its owner.

Platform Bed

This massive platform bed was designed specifically to suit the exotic tastes of a special client. Made from solid African bubinga and African wenge, the piece is as heavy as it is beautiful. Because of the flat orientation of the side supports, there were numerous challenges to overcome to fully support the weight of a queen-size bed and its occupants. In the end, the bed appears to be floating with no visible center supports.

Because the workpieces were so large, this project served as a great opportunity to use hybrid techniques: power tools for the grunt work and hand tools for the finessing.

Level the End Grain ▪ The block plane levels exposed end grain in the glued-up footboard.

Level the Long Grain ▪ The top of the footboard also needs a little leveling after the glue-up.

The Footboard

To achieve the desired thickness, the massive footboard was made from three pieces of solid 8/4 bubinga. Getting everything to line up perfectly with pieces of this size can be a challenge. So I got as close as I could using the Festool Domino for alignment help. After the glue dried, the slight discrepancies in the end grain were cleaned up using a block plane. Any unevenness in the top of the footboard was leveled with the block plane and a scraper.

Wenge Panel

The headboard of the bed features a raised panel made of solid wenge. This solid panel was made up of two boards glued together. Big panel glue-ups are rarely, if ever, perfect and need subsequent finessing. A finely sharpened No. 80 cabinet scraper does a fine job of removing the glue, leveling the surfaces and keeping the panel as close to flat as possible.

Wenge Headboard ▪ The wenge headboard is glued up from two boards.

Scrape the Glue Line ▪ The cabinet scraper cleans and levels the joint.

Big Mortise-and-Tenon Joints

Large mortise-and-tenon joints hold the long side supports into the headboard and footboard. Even though this bed features knock-down hardware, these joints support all the weight of the bed.

The mortise was made in several passes using a ½" spiral bit in a plunge router. While I usually round over the tenon instead of squaring up the mortise, the wavy ends of this particularly large mortise would interfere with the rounded tenon. Squaring off the mortise is the only logical choice. I chopped it square using a large mortise chisel – a good match for the bubinga, a dense tropical hardwood.

I scored the tenon's shoulder lines using a cutting gauge. At the table saw, with the workpiece securely held on the crosscut sled, the dado stack removed the bulk of the material. Back at the workbench, I finessed the fit with the rabbeting block plane.

Big Mortise ▪ The large headboard mortises were routed.

Square 'Em Up ▪ The wavy ends of the mortise need to be squared off.

Big Tenons ▪ Scribe the tenon shoulders with a cutting gauge.

Table Saw Removes the Bulk ▪ Using a crosscut sled and a dado stack, the tenons are cut to rough size at the table saw.

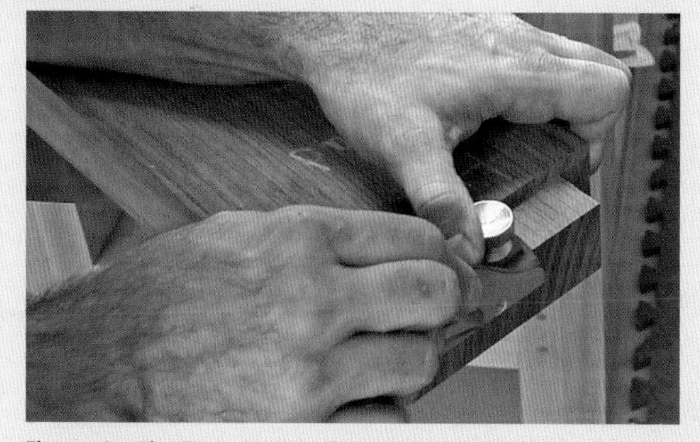

Finessed to Fit ▪ The rabbeting block plane smooths the tenon cheek.

Split-Top Roubo Workbench

A Monster Workbench ■ The split-top Roubo-style workbench has everything a woodworker needs, whether you use power tools, hand tools or both.

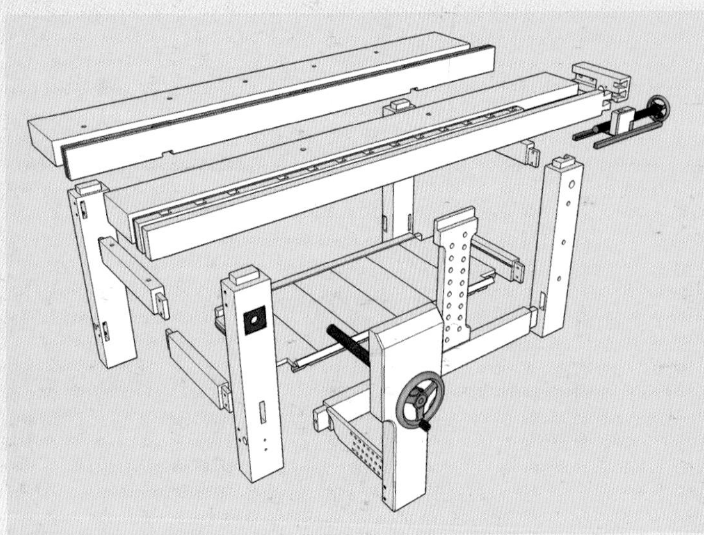

The split-top Roubo workbench is a modern twist on the classic Roubo bench. Inspired by Plate 11 of André Roubo's *L'Art du Menuisier* (1769), Benchcrafted modified the design to complement their unique vise hardware. The split-top Roubo is a proven design that will serve as a lifetime workbench and will never need to be replaced. Whether you're into hand tools, power tools or both, a bench like this will serve all of your workholding needs.

Big Band Sawn Dovetails

The workbench features massive half-blind dovetails where the end cap meets the front strip. Cutting them by hand would be quite difficult, but the band sawn dovetail method (page 154) makes an otherwise difficult task simple and accurate.

I laid out the tails first, with the bevel gauge set to seven degrees. At the band saw, the tapering jig guided the angled tail cuts. After removing the waste and handsawing the shoulders, I chiseled the shoulders flat and clean.

The pin locations on the end cap could then be marked from the tail board and routed out. Here's a clever trick I learned from Jameel Abraham of Benchcrafted: Start by routing a ¼"-deep socket, then chisel its walls clean. A bearing-guided pattern bit now can follow that shallow socket to remove the rest of the waste.

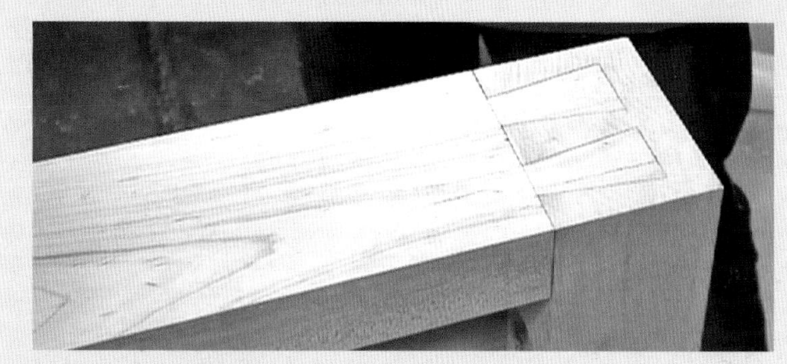

Condor Tails ▪ The large dovetails in the front apron would be tricky to make without the help of power tools.

Layout First ▪ Lay out the dovetails using a sharp pencil and a bevel gauge.

Saw the Tails ▪ At the band saw, an angle fixture holds the work at the dovetail angle.

Handsaw is Best ▪ With pieces this size, it's easier to handsaw some of the rough cuts.

Cleanup with a Chisel ▪ A sharp chisel cleans up the shoulder of the dovetail joint.

Transfer the Tails ▪ Trace the tails onto the end cap with a sharp pencil.

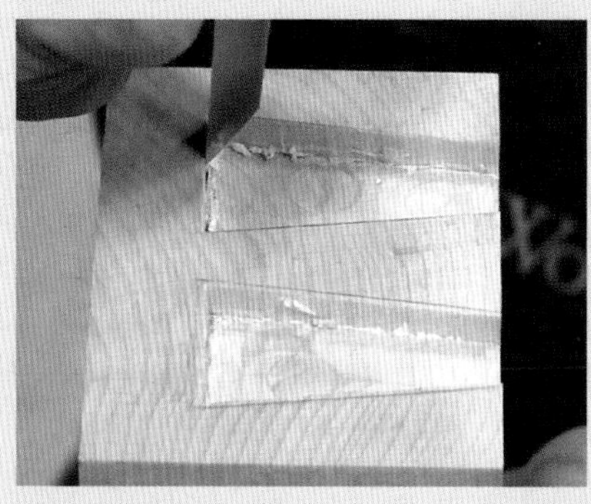

Rout ¼" Deep ▪ Rout the pin socket to ¼" deep and clean it up with a chisel to prepare for the pattern bit.

Wall-Hanging Cabinet

Wall-Hanging Cabinet • An exercise in elegant casework.

The wall-hanging cabinet is an exercise in basic case construction as well as door and drawer construction and fitting. I made this cabinet in 2011. I built two versions, one with plywood and another with solid wood. The solid-wood version features a dovetailed case and drawer along with mortise-and-tenon joints for the frame-and-panel doors. The doors were attached using elegant knife hinges. The plywood version features standard dado and rabbet joints for the case and the drawer. The doors were assembled using pocket screws and attached to the case with no-mortise hinges.

Building two similar but different cabinets gave me the opportunity to demonstrate numerous hybrid techniques that were applicable to a wide range of budgets, skill levels, materials and tastes.

Stopped Dados

The solid-wood case features a series of dados for the middle and bottom shelves. To look good, the dado must stop before it reaches the front of the case. I routed the stopped dado using an adjustable-width dado jig and a straight bit.

To square up the dado's rounded end, I referenced a sharp chisel off the existing dado walls and extended them to the end point. With both side walls scored, the bulk of the waste can be removed by working back to the layout lines. Only after the extra material is out of the way is it safe to chop right on the layout lines.

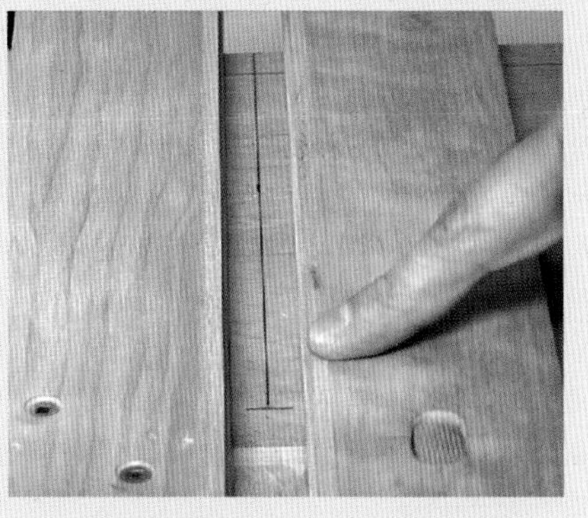

Dados with a Router Jig • An exact-width dado jig makes it quick to rout the dados.

Stopped Dados • Square up the stopped dado by extending the side walls.

Establish the End • Using the layout lines as a guide, chop the end of the stopped dado.

Nice and Square • This stopped dado is squared up and ready to receive the shelf.

Knife Hinges

Knife hinges may look complicated, but they are easy to install once you know the sequence of steps. It's a lot like mortising for a butt hinge (page 149). Begin on the case and outline the hinge with a marking knife. Then rout the bulk of the material to the exact depth for a flush fit, taking care to stay away from the lines.

Remove the excess material with a sharp chisel. To ensure a snug fit, make the final strokes with the chisel sitting directly in the knife line. Mortise the door the same way for the other half of the hinge. The result is a smooth-moving door that swings freely.

Scribe the Outline • Using a sharp marking knife, scribe the perimeter of the hinge leaf.

Work to the Scribe Lines • After routing, chisel back to the scribe lines.

A Snug Fit • The hinge should rest snugly in the mortise.

New Spin on a Summertime Classic ■ The Adirondack chair gets a Greene & Greene makeover, which presents numerous opportunities and challenges for the hybrid woodworker.

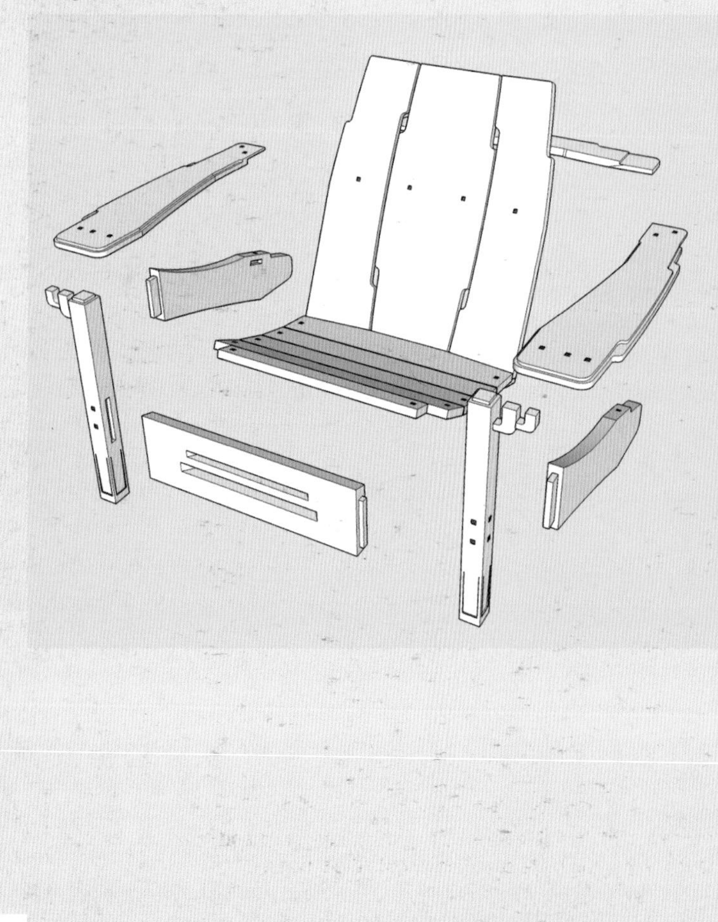

Adirondack Chair

This Adirondack chair features classic Greene & Greene styling. It was built for looks as well as comfort. Small details including the repeated cloudlifts, ebony plugs and leg indents give it a unique look and provide numerous challenges and opportunities to build woodworking skills. I am a huge fan of Greene & Greene furniture and I wanted to bring in elements from my favorite pieces. So the challenge was to do this in a way that would be accessible to the average woodworker and I believe we accomplished that. There are some unusual mortise-and-tenon joints in this chair, which I couldn't have made without hybrid woodworking techniques.

Template

The template for this project was cleverly designed with the help of my friend Aaron Marshall. We came up with a simple two-sided template that could be used to create every cloudlift and profile in the chair. The template itself was made from particle board, which is not my favorite material, but it was available and cheap. A paper template was glued to the surface of the template stock and the shape was rough-cut at the band saw. To finesse the final shape of each profile, I used rasps and worked my way back to the paper template guide lines.

The template was then used to trace the appropriate shapes onto the arms and back pieces. After rough-cutting at the band saw, each piece was carefully finessed using rasps and a flexible sanding strip.

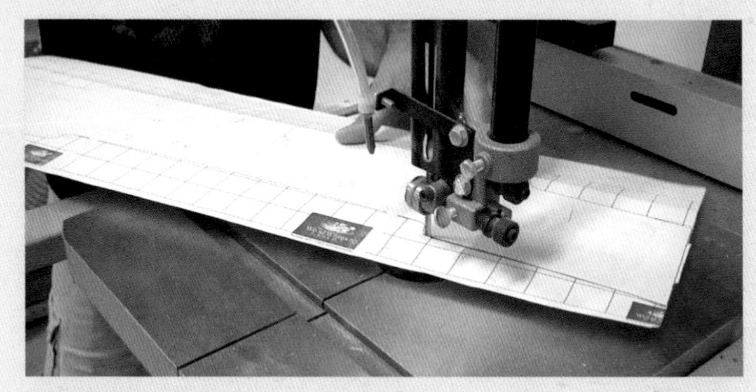

Rough-Cut the Template ■ Using a printout as a guide, the template is rough cut at the band saw.

Refine the Shape ■ A rasp brings the template to its final shape.

A Shallow Mortise ■ The shallow mortise under the chair arms was routed, then squared with a chisel.

Stubby Mortise-and-Tenon Joints

Where the front legs meet the arm rests, there is a stubby mortise-and-tenon joint. The shallow mortise was routed into the underside of each arm using a straight bit. The mortise was then cleaned up and squared off with chisels.

The short tenons at the top of the front legs were cut at the table saw with the miter gauge. To ensure crisp shoulders, a cutting gauge was used to establish the shoulder lines prior to sawing. Because the tenon is so short, a chisel was the best tool for fitting the tenon cheeks.

Scribe the Shoulders ■ The cutting gauge marks the shoulders of the short front-leg tenon.

Quick Work at the Table Saw ■ Using the miter gauge and the rip fence, the table saw quickly cuts the tenons.

Side to Front Mortise-and-Tenon Joints

The mortises in the front legs were laid out using a pencil and an adjustable square, and then cut out using a plunge router and an edge guide. The legs for this chair are wide enough to safely support the router though the entire cut, so there's no need for auxiliary support.

Where the side leg meets the front leg, a strong mortise-and-tenon joint is essential. Because the side leg approached the vertical front leg at an angle, the tenon shoulder had to be cut at an angle on the table saw with a miter gauge. With the shoulder established, the tenon required additional work best done with hand tools. The short shoulders of the tenon were established using a handsaw, then the flush-trim saw cut the proud shoulder flush. The shoulder was finished off with a wide chisel. The tenon was then rounded to fit the round-end mortises.

Angled Tenons ■ A table saw and a precise miter gauge is a great way to make the initial cuts for angled tenons.

Easier with a Handsaw ■ When it comes to angled tenons, sometimes it's easier to use a handsaw.

Flush-Trim the Shoulders ■ Remove the proud shoulder material using a flush-trim saw.

Pare What's Left ■ Finish the shoulder by paring with a sharp chisel.

Leg Curves

The side legs have two simple curves created with a drawing bow. The curve was then rough-cut on the band saw and finessed with a spokeshave and some sanding. When one leg's curves were complete, I used double-stick tape to attach the two legs together so a flush-trim bit could make the second leg match perfectly.

Lay Out the Curves ■ A drawing bow lays out the large curve on the side legs.

Refine With a Spokeshave ■ A spokeshave refines the curve.

Flush-Trim the Second Leg ■ Using the first leg as the template, trim the second leg to shape on the router table.

Leg Detail

The front legs receive several elegant details, such as a slight rounding over of the leg bottom. A small template transferred the desired shape to each side of the leg. A block plane then removed the wood back to the pencil lines on all four sides of each leg. This slight roundover gives the leg a light appearance.

A Nice Detail ▪ A small template transfers the desired shape to the bottom of each leg.

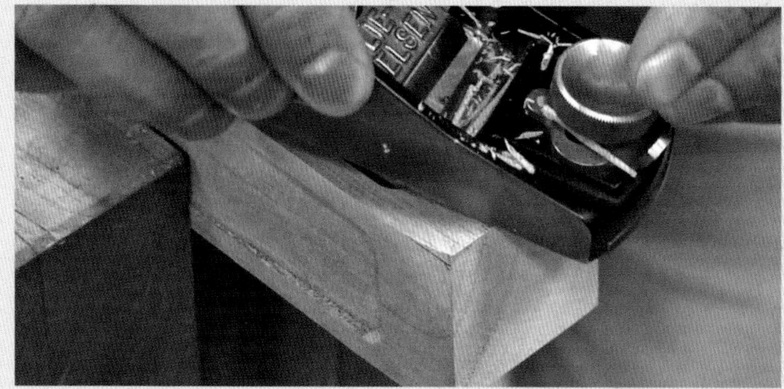

Plane to the Line ▪ A block plane makes quick work of removing the wood to the lines on the leg.

Front Apron Detail ▪ The slots in the front apron were routed with an edge guide.

A Gradual Ramp ▪ Saw the side walls to form the ramp that ends each slot.

Front Apron Slots

Two through slots decorate the front apron of the chair, a design feature borrowed from the Greene & Greene Gamble House dining table. I used a router with a straight bit and an edge guide to rough out the slots.

At the end of each slot, a ramp needs to be created at about 45 degrees. Using a handsaw, the sides of the ramp are roughly established. With a chisel, the excess material can be removed between the saw kerfs. The final finessing is done using a rasp as the 45-degree ramp becomes a graceful profile.

Chisel Away ▪ Remove the wood between the walls with a chisel.

Refine with a Rasp ▪ A small rasp smooths and refines the ramp.

Marc's Last Word

I must admit, I love my power tools. So much so that I have earned a bit of a reputation as a power-tool junkie. But as you can tell by now, I also have come to enjoy an additional sense of pride and craftsmanship that comes with traditional hand tools. So the truth is, I'm not a power-tool junkie, I'm a woodworking junkie. I love experimenting with new tools and techniques with the hopes of streamlining my process for efficiency, accuracy and, yes, enjoyment.

Whenever big-picture discussions about tool choice and usage comes up, there's always at least one person who reminds everyone that the recipient of the work could care less what tools you used to make it. That's absolutely true and many times, I'm the guy saying that. But hand-crafting furniture, for many of us, is as much about the process as it is about the product. Although it is intangible, the sense of pride we feel about our projects and the love and care that goes into their creation is the stuff that makes them unique and desirable.

I'm a big fan of sushi. I'm certainly not a connoisseur but I do enjoy eating it. While I really don't care what tools the chef uses to prepare the meal, I certainly can tell the difference between sushi made by someone who is skilled and passionate and sushi made by someone who is only there to collect a paycheck. Sometimes there's a noticeable quality difference and sometimes it's an intangible element that I can't quite put my finger on. But a passionate sushi chef puts a bit of himself into every piece he makes, and woodworkers should do the same. My furniture is more valuable simply because I care. I care about everything from the raw boards it was made from to the tools I used to make it.

This is why I feel exploring tools and methods is a good use of your time. It will help you refine your woodworking while also helping you to derive more pleasure from the experience. If that exploration lands you in a 100 percent power-tool shop or a 100 percent hand-tool shop, then you are exactly where you're supposed to be. But I would guess that the average woodworker is going to find a home somewhere in between the extremes in what might be considered the largest and most popular category that I refer to as hybrid woodworking.

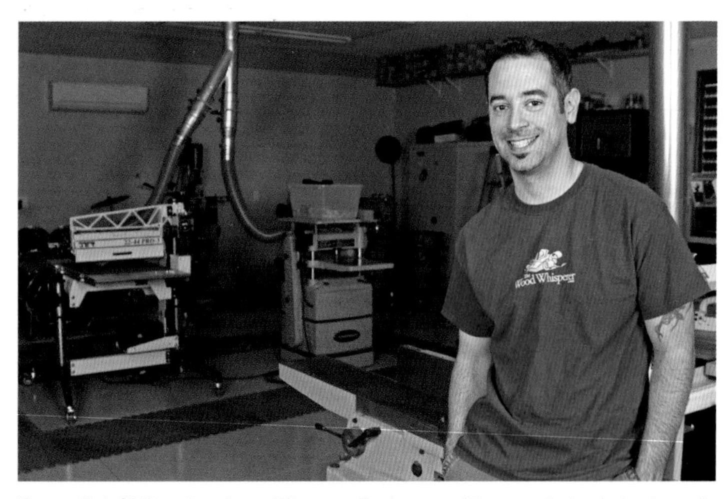

Happy Hybrid Woodworker ▪ When you're happy with your choice in tools and techniques, it's easy to have pride in your work.

Where to go from here?

I hope you'll head to the workshop and try the hybrid approach for yourself. You'll need to experience the increased control, better accuracy and higher levels of gratification for yourself before you can be sold on the concept. If you find some of the must-have tools missing from your shop, don't go on a shopping spree. Instead, incorporate one tool at a time as you take on the hybrid way of thinking, and eventually you'll have everything you need.

Once the system makes sense to you, it's time to explore. There are so many tools out there and exponentially more techniques to consider. Learn from as many woodworkers as you can and always keep an open mind. The minute you are convinced you absolutely have the best way of doing something is the minute you stop learning about it. Instead, continue being a lifelong student of the craft, and custom design the woodworking journey that feels right to you.

Hybrid Woodworking.
 Published by Popular Woodworking Books, an imprint of F+W Media, Inc., 10151 Carver Rd., Suite 200, Blue Ash, Ohio, 45236. (800) 289-0963 First edition.

Distributed in Canada by Fraser Direct
100 Armstrong Avenue
Georgetown, Ontario L7G 5S4
Canada

Distributed in the U.K. and Europe by
F&W Media International, LTD
Brunel House, Ford Close
Newton Abbot
TQ12 4PU, UK
Tel: (+44) 1626 323200
Fax: (+44) 1626 323319
E-mail: enquiries@fwmedia.com

Distributed in Australia by Capricorn Link
P.O. Box 704, Windsor, NSW 2756 Australia
Tel: (02) 4560 1600; Fax: (02) 4577 5288
Email: books@capricornlink.com.au

Visit our website at popularwoodworking.com or our consumer website at shopwoodworking.com for more woodworking information projects.

Other fine Popular Woodworking Books are available from your local bookstore or direct from the publisher.

ISBN-13: 978-1-4403-2960-9

17 16 15 14 13 5 4 3 2 1

Acquisitions editor: David Thiel
Content editor: John Kelsey
Copy editor: Megan Fitzpatrick
Designer: Daniel T. Pessell
Production coordinator: Debbie Thomas

Read This Important Safety Notice

To prevent accidents, keep safety in mind while you work. Use the safety guards installed on power equipment; they are for your protection.

When working on power equipment, keep fingers away from saw blades, wear safety goggles to prevent injuries from flying wood chips and sawdust, wear hearing protection and consider installing a dust vacuum to reduce the amount of airborne sawdust in your woodshop.

Don't wear loose clothing, such as neckties or shirts with loose sleeves, or jewelry, such as rings, necklaces or bracelets, when working on power equipment. Tie back long hair to prevent it from getting caught in your equipment.

People who are sensitive to certain chemicals should check the chemical content of any product before using it.

Due to the variability of local conditions, construction materials, skill levels, etc., neither the author nor Popular Woodworking Books assumes any responsibility for any accidents, injuries, damages or other losses incurred resulting from the material presented in this book.

The authors and editors who compiled this book have tried to make the contents as accurate and correct as possible. Plans, illustrations, photographs and text have been carefully checked. All instructions, plans and projects should be carefully read, studied and understood before beginning construction.

Prices listed for supplies and equipment were current at the time of publication and are subject to change.